W0256464

Dr. sc. nat. Armin Hemmerling

Born 1948 at Grambow near Schwerin. Studied mathematics at E. M. Arndt
University, Greifswald, from 1966 to 70; research student of mathematical cybernetics from 70 to 73. He was with the Dept. of Mathematics
of E. M. Arndt University, where he received the Dr. rer. nat. degree
in 74 and the Dr. sc. nat. degree in 81. Since 1988 he is with the
Dept. of Mathematics and Physics of the Pädagogische Hochschule "Li-
selotte Herrmann", Güstrow.
Fields of interest: Computation theory, algorithmic complexity.

Hemmerling, Armin:
Labyrinth Problems: Labyrinth-Searching Abilities of Automata /
Armin Hemmerling. - 1. Aufl. - Leipzig : BSB Teubner, 1989. -
215 S.
(Teubner-Texte zur Mathematik ; 114)
NE: GT

ISBN 978-3-322-94561-7   ISBN 978-3-322-94560-0 (eBook)
DOI 10.1007/978-3-322-94560-0

ISSN 0138-502X

Gesamtherstellung: Typodruck Döbeln, Bereich Leisnig
Bestell-Nr. 666 535 8
02700

Armin Hemmerling

Labyrinth Problems

Labyrinth-Searching Abilities of Automata

This book deals with the main results and open problems of labyrinth theory which has to be considered as a branch of complexity theory within theoretical computer science. The central problems treated here concern the searching of certain embedded graphs by several kinds of automata able to move in these graphs.

Dieser Teubner-Text behandelt die wichtigsten Resultate und offenen
Probleme der Labyrinth-Theorie, die als Zweig der Kompliziertheits-
theorie innerhalb der theoretischen Informatik anzusehen ist. Im Zen-
trum der Untersuchungen stehen hier Probleme des Absuchens gewisser
eingebetteter Graphen durch verschiedene Typen von Automaten, die in
der Lage sind, sich auf diesen Graphen zu bewegen.

Ce livre est consacre au traitement des résultats les plus importants
de la théorie des labyrinthes et à celui des problèmes qui restent
ouvert. Cette théorie est ici considérée une branche de la théorie de
la complexité en informatique théorique. Au centre de ces recherches
se trouve le problème de l'explorations de certains graphes immerges
par différents types d'automates qui sont en mesure de se mouvoir sur
ces graphes.

Книга содержит главные результаты, а также нерешённые проблемы теории
лабиринтов, которая рассматривается как ветвь теории сложности в об-
ласти теоретической информатики. При этом, в центре исследований стоят
проблемы обхода некоторых погруженных графов, посредствам различных
видов автоматов, обладающих способностью двигаться по этим графам.

CONTENTS

<u>Acknowledgements</u>

I would like to thank the following people who have contributed
to this book in several ways : C. Bandt , I. Bookhold , L. Budach ,
M. Bull , M. Ejsmont , F. Hoffmann , K. Kriegel , Joh. Lehmann ,
H. Müller , H. Sachs , P. Schreiber , J. Ulehla , L. Voelkel ,
G. Wechsung.
I am particularly indepted to Prof. Dr. G. Asser for encouraging
me to write a monograph on labyrinth theory and for numerous
inspiring discussions on this subject.
Dr. W. Schleinitz read provisional versions of the manuscript
with great care and made a lot of vital comments,especially con-
cerning the English style.
Furthermore,I want to thank the colleagues from the staff of the
Teubner-Verlag for the cooperation.
Finally,I am most grateful to Elke , Kathi and Robert for toler-
ating my time-consuming solitary travels through the metalabyrinth
of labyrinth research.

February 1989   A. H.

> Denn wie jede große Idee hat es
> eigentlich keinen Anfang,sondern ist,
> eben der Idee nach,immer dagewesen.
>
> Herrmann Hesse, Das Glasperlenspiel

INTRODUCTION

Already since the earliest eras of civilization,the idea of
mazes or labyrinths has been well-known,and they have often been
used to allegorize complicated or mysterious things and hardly
solvable problems.

In the ancient Greek mythology the labyrinth of Minos,who was
the king of Crete,is mentioned. In this labyrinth a terrible monster lived,the Minotaur. It fed on human flesh. Every ninth year,
seven Athenean young men and virgins had to be sacrificed to the
monster. Fortunately,in the third victimization,Theseus,the
prince of Athens,killed the Minotaur. After that,as the legend
tells,Theseus found the way out of the labyrinth by means of a
thread of wool which he had used to mark his way when walking
through the labyrinth searching for the Minotaur.

This story illustrates two basic problems connected with labyrinths. One problem is to find an exit,starting from an arbitrary
position in the labyrinth. This task had to be solved by Theseus
after his victory over the Minotaur and must be solved by any
adventurer who has lost his way in a labyrinth,if he wants to
return to the outer world. But the problem Theseus had to solve
first (and which does not seem to be noted in the myth) was to
search the whole labyrinth in order to find the domicile of the
Minotaur.

We suppose that Theseus or the adventurer mentioned above do
not know the plan of the labyrinth. They may at most know that it
is of a certain type,for instance,that it is arranged in a plane.
Therefore,they have to use procedures which guarantee the success
in any labyrinth of the corresponding type.

We summarize the basic problems by provisional formulations.

<u>Searching problem.</u> Give a procedure for searching every laby-
rinth of a certain type,i.e. eventually to reach any point
of the labyrinth,starting from an arbitrary position.

5

<u>Escaping problem.</u> Give a procedure for escaping from every laby-
     rinth of a certain type,i.e. eventually to reach an exit,
     starting from an arbitrary position.

   Obviously,any solution of the first problem can be used to
solve the second one. Of course,also the time needed for searching
and escaping,respectively,plays an important role in the evalu-
ation of the procedures solving these problems.

   For the time being,by a labyrinth we mean a complex consisting
of some rooms and some corridors which connect these rooms. More
precisely,a corridor can connect two rooms or a room with an exit
of the labyrinth. Any two corridors must not intersect each other.

   Arriving at some room,a person who has to walk through the
labyrinth can only see the corridors terminating at this room and
the markers which he has possibly attached to the room or to some
of these corridors when he entered them at some earlier time.
Depending on this local information,the person must decide through
which of the corridors he has to move now.

   This myopic behaviour corresponding to the topology of the
given object is a basic characteristic of labyrinth problems and
will be referred to as the <u>inner</u> <u>point</u> <u>of</u> <u>view</u> in the sequel. In
contrast to this,the outer point of view would characterize the
search for an exit with the finger on the map of the labyrinth,
possibly regardless of its topology.

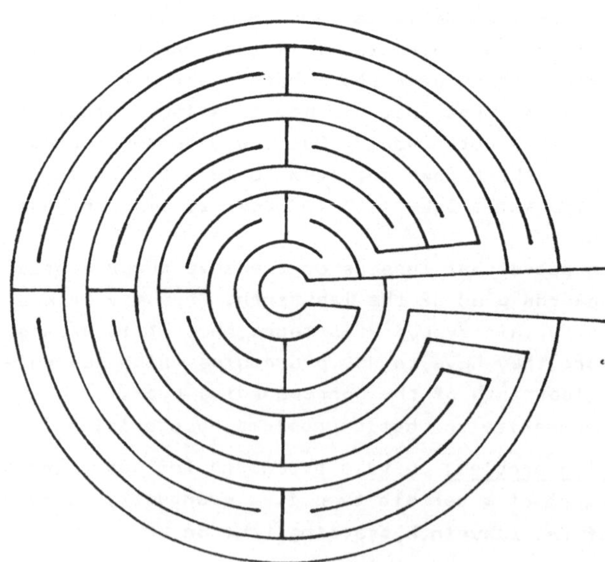

Fig. 1

We would like to remark that many descriptions and illustrations of the labyrinth of Crete and of other mazes show that the
authors did not really understand the difficulties of labyrinth
problems. For example,the Roman poet Ovid compared the Cretan
labyrinth with the winding course of the river Meander,cf. the
quotation at the head of Chapter III. In ancient illustrations
a labyrinth is often represented as a single,frequently winding
corridor. Fig. 1 gives the map of the Cretan labyrinth as it is
shown in an Italian engraving,see /S.Ma/. Of course,labyrinths of
this kind can easily be searched. The difficulties of labyrinth
problems are caused by branchings.

But there were real labyrinths,too,e.g. the networks of the
Roman catacombs or the dungeons and subterranean passages in medi-
eval castles. Recently U. Eco used a labyrinth as the scene of
some thrilling situations in his bestseller (see quotation at the
head of Chapter II). Unfortunately,his hero was not able to master
this simple labyrinth,shown in Fig. 2,from the inner point of
view.

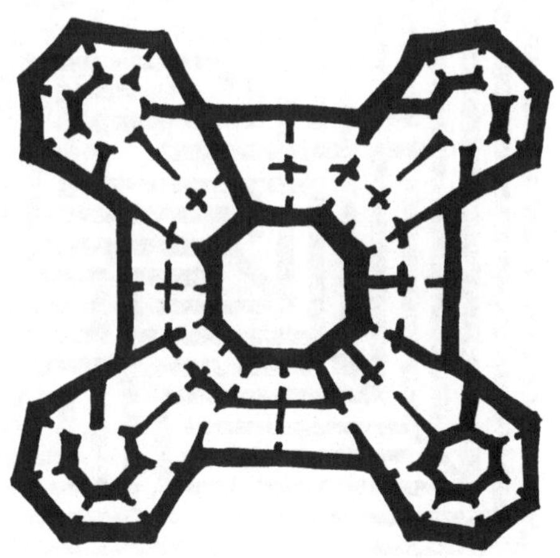

Fig. 2

In a much more harmless way the garden architecture,especially
of rococo,was enriched by the idea of labyrinths. We show a plan
of the famous maze in the garden of Hampton Court (London; Fig. 3)
and the labyrinth in the park of the village Altjeßnitz (near
Wolfen,G.D.R.; Fig. 4).

Finally,we mention the many nice labyrinths given by our modern
cities. The reader may take a city map of his or her home town
as an example.

Fig. 4

The first mathematical publication on labyrinth problems was by C. Wiener /L.Wi/, who gave a solution of the searching problem for finite graphs. Other **and simpler** procedures for solving this problem were presented by Tarry and **Trémaux**. D. König surveyed these results in a chapter of his **famous** book on graph theory /L.Ko/.

Whereas König's concept of (undirected) graph gave a provisional definition of the objects which have to be searched,

8

C. Shannon /L.Shan/ contributed the idea of the subjects or devices having to walk in these objects. According to him,automata of a certain type,so-called mice,were put into labyrinths in order to search them or to find exits. This was the origin of modern labyrinth research within the framework of computer science.

Subsequently,many authors have contributed results and applications,but also open problems and motivations to this area. We sketch some details.

In the sixties there were some papers considering the classic algorithms or slight modifications of them within the framework of automata theory or formal language theory,/L.KoTh/,/L.Sc/, /L.Ros/,/S.KoTr/.

M. Blum and A. Rosenfeld and their co-workers have considered labyrinth problems in connection with digital image processing and pattern recognition,/S.BlHe/,/L.BlSa/,/L.BlKo/,/S.My/, /S.Ro76/. Indeed,for any information processing with non-linear inputs,such as digital images or patterns,the problem of searching (traversing) the inputs is non-trivial. On the other hand,it has a great importance as a basis for decision processes and other more complicated tasks.

For the same reason,searching algorithms are useful in programming cellular networks of automata /S.RoFiHo/ or in comparing cellular and iterative arrays with Turing machines with multidimensional tapes /S.He79a,b,86/.

The configurations of a nondeterministic (space-bounded) Turing machine with some fixed input word form a finite directed graph in a natural sense. In order to recognize whether the input is accepted by some computation,it would be sufficient to determine whether this configuration graph is traversable or not,i.e., whether an accepting configuration is reachable if one starts with the initial configuration. In this way,W. Savitch /S.Sav70, 73/ showed the NL-completeness of a certain traversability problem. Unfortunately,his labyrinths are directed graphs,whereas in usual labyrinth research undirected graphs have preferably been considered so far. The main reason that those papers do not belong to labyrinth theory in the strict sense is,however,that Savitch's maze-recognizing automata do not work according to the inner point of view. Indeed,their moves do not necessarily correspond to the topology of the mazes they are working on. Contrary to this,the papers /L.AlKaLiLoRa/ and /L.CoRa/ which are caused by Savitch's result consider proper labyrinth problems.

Another connection between searching problems for certain undirected graphs,namely multidimensional mazes,and the problem of

determinism versus nondeterminism for the space complexity is
given in /S.HeMu/.

In Europe modern labyrinth research was stimulated by K. Döpp
/L.Do/. His conjecture that there is no finite automaton which
masters every co-finite plane maze (Rosenfeld /S.Ro76/ asked the
related question if there is a finite automaton searching every
finite plane maze) was confirmed by L. Budach /L.Bu75,78a,b/. The
corresponding theorem is a milestone of labyrinth theory. Budach
developed a general conception for investigations of the behaviour of automata in environments,in order to apply the results to
arbitrary computation models,/S.Bu/,/S.BuMe/.

Besides these motivating or applicable results,some more or
less internal advances within labyrinth theory could be observed.
We only mention the nice idea of the construction of traps for
cooperating automata,which was developed by M. Rabin (unpublished)
and used in /L.Ro/ and /L.He87a/.

Summarizing one can say that several interesting and useful
results have been produced about automata in labyrinths,but many
important problems have not yet been solved. Some of these open
problems were known for years and were attacked by well-known
researchers. But whereas in the seventies the investigation of
labyrinth problems was quite up-to-date,an important problem of
today is the search for theorists working in this area. On the
other hand,except that labyrinth problems are connected with essential questions of modern computer science,the idea of labyrinth
is present in a great variety of solitaire games and puzzles,
computer programs and even in belletristic literature. Therefore,
it is appropriate to give a survey of the present state of theoretical research.

The present book deals with the main results,tecnniques and
open questions of labyrinth research. Labyrinth theory can be
considered as both a branch of and an alternative to complexity
theory. Indeed , it tries to characterize the algorithmic complexity of certain problems. This includes the static (or descrip-
tional) complexity characterized by the types and the numbers of
internal states of automata performing certain tasks , as well as
the dynamic (or computational) complexity estimated by lower and
upper time and space bounds for the tasks. On the other hand,laby-
rinth theory has its own characteristic area of problems which
preferably concern the moves of automata in certain non-linear
structures in order to traverse (search or master) them or to
recognize their properties. So it yields a new reservoir of

examples and subjects for studying the abilities of abstract
machines.

The choice of material for the presentation in this book is
devoted to the inner point of view in a twofold meaning. First,
we restrict ourselves to such types of labyrinths and automata
which do not violate the inner point of view as it has been explained above. This is the characteristic of labyrinth theory in
the narrower sense. Secondly,we consider only the hard core of
labyrinth theory and do not deal with applications of labyrinth
results to other branches of complexity theory or computer
science. Such a restriction is necessary because of the space
limitations. Moreover,it seems to be justified and even useful
in order to stress that our topic is relatively selfcontained.

Finally,with respect to the subject of labyrinth theory,one
should remark that the former meaning of the inner point of view
also includes that the objects are given by themselves,but not
in a coded form. In contrast to this,estimations of the complexities of properties or tasks within P-NP theory /S.GaJo/ or computational geometry /S.PrSh/ are based on more or less natural
codes of the objects. Also the searching problems in abstract
data structures,see /S.Wi/,are beyond the scope of labyrinth
theory.

This book is organized as follows. Chapter I deals with the
basic concepts and facts on labyrinths,automata working in these,
and problems considered in labyrinth theory. In Sections 1.7 -
1.10 , we introduce some notations and techniques which illustrate
the basic concepts,but are mainly fundamental for trap constructions. The reader who is preferably interested in searching
methods should switch from Section 1.6 to Chapter II.

Chapter II presents searching algorithms and,therefore,upper
bounds of the complexities of searching problems. These algorithms
are mainly described by certain programs,in some cases they are
given by informal descriptions of their behaviour.

Chapter III is devoted to trap constructions as far as they
are representable on a few pages. In this way,lower bounds of the
complexities of searching problems are fixed.

Finally,in Chapter IV,we try to complete the survey of labyrinth theory by stating some supplementary results without proofs
and by sketching some open problems.

All sections of Chapters I - III are closed with hints to the
sources of the presented results and concepts or to related facts
and problems. The notations introduced in this book are underlined

where thev are defined. By the index,one can find the pages of
definitions and results ; moreover,the used symbols are listed.
Within the text,by Theorem X we refer to Theorem X in the current
chapter , whereas Theorem Y.X means Theorem X in Chapter Y. Correspondingly we refer to propositions,corollaries,lemmas and
figures. The end of a verification of a result is marked by //.
Note that the bibliography consists of three parts indicated by
L.,S. or Q. in the references.

CHAPTER I. BASIC CONCEPTS

In this chapter we introduce the basic types of labyrinths, the automata able to operate on those,and the problems we shall deal with in the sequel. Moreover,some fundamental facts concerning these concepts are summarized.

1.1. Graphoids

Although the concept of graph is the essential basis for defining labyrinths,from technical reasons it is useful to go back to objects whose elements are vertices and half-edges instead of vertices and edges in graphs. Such objects are called graphoids in the sequel. This name seems to be still nearly vacant. We only found a mention (however with another meaning) in /S.Ha/,p. 41.

A _graphoid_ is a triple

$$G = (V,H,I),$$

where

- V is a non-empty set (whose elements are called _vertices_),
- H is a set disjoint to V (its elements are the _half-edges_ of G), and
- I is an irreflexive symmetric binary relation in $V \cup H$ such that the following conditions are satisfied.
  - (1) $I \cap V^2 = \emptyset$;
  - (2) for each $h \in H$ there is exactly one $v \in V$ with $(h,v) \in I$;
  - (3) for each $h \in H$ there is at most one $h' \in H$ with $(h,h') \in I$.

The vertices and half-edges are also called _elements_ of the graphoid G. I is the _incidence relation_ of G,and two elements are said to be _incident_ if they are in relation I one to the other. Let H(v) denote the set of all half-edges incident to vertex v.

The vertex incident to the half-edge h is denoted by ver(h),and
hal(h) is the half-edge incident to h if there is any.

A half-edge h is said to be <u>free</u> if it is not incident to
another half-edge,i.e.,if hal(h) is not defined; otherwise it is
called <u>bound</u>. By an <u>edge</u> of the graphoid G,we mean a set

$$e = \{h,h'\}$$

consisting of two mutually incident half-edges h and h'. We shall
say that the edge e is incident to the vertices ver(h) and
ver(h'). Besides these undirected edges,sometimes we also have to
consider the <u>directed edges</u>,those are the pairs $(h,h') \in I \wedge H^2$.

A graphoid G is said to be <u>open</u> if it possesses free half-
edges. If all half-edges are bound,it is called a <u>graph</u>. In this
case let

$$E = \left\{ \{h,h'\} : \ (h,h') \in I \wedge H^2 \right\},$$

this is the set of all edges,and

$$f(\{h,h'\}) = \begin{cases} \{v,v'\} & \text{if } v = \mathrm{ver}(h) \neq \mathrm{ver}(h') = v', \\ \{v\} & \text{if } v = \mathrm{ver}(h) = \mathrm{ver}(h'). \end{cases}$$

Then the triple (V,E,f) is an undirected graph according to a
suitable definition of this concept allowing multiple edges and
loops,cf. /S.Be/,/S.Or/ or /S.Sa/. Here f is the incidence function assigning to any edge the set of its endvertices.

A graphoid is said to be <u>finite</u> if it has only finitely many
elements. Sometimes we also have to consider graphoids with infinitely (but denumerably) many elements.

A graphoid G is called <u>simple</u> if it does not possess loops or
multiple edges. This means $\mathrm{ver}(h) \neq \mathrm{ver}(h')$ for any edge $\{h,h'\}$,
and $\{\mathrm{ver}(h_1),\mathrm{ver}(h_1')\} \neq \{\mathrm{ver}(h_2),\mathrm{ver}(h_2')\}$ for any two edges
$\{h_1,h_1'\} \neq \{h_2,h_2'\}$.

Often a graphoid will be represented or even defined by a
figure. Then the vertices will be drawn as dots and the half-
edges as smooth lines such that the incidence relation is represented by the coincidence of endpoints of lines with dots and by
the coincidence of endpoints of lines,respectively. Divisions of
the edges into half-edges will not be specified,since they are
not essential for our purposes. The free ends of half-edges will
be represented by small semicircles.

For instance,Fig. 1a shows a finite open graphoid,whereas by
Fig. 1b an infinite graph is represented. In the latter,the three
small dots indicate the continuation ad infinitum.

Naturally,by an <u>isomorphism</u> between the graphoid G and a
graphoid G' = (V',H',I') ,we mean a bijection φ of $V \cup H$ onto
$V' \cup H'$ such that

$$\varphi('V) = V' \ , \quad \varphi('H) = H' \ ,$$

and

$$(g_1,g_2) \in I \quad \text{iff} \quad (\ \varphi(g_1), \varphi(g_2)\ ) \in I'$$

for any two elements g_1,g_2 of G. In this case, G and G' are said to be _isomorphic_, or G' is called an _isomorphic copy_ of G.

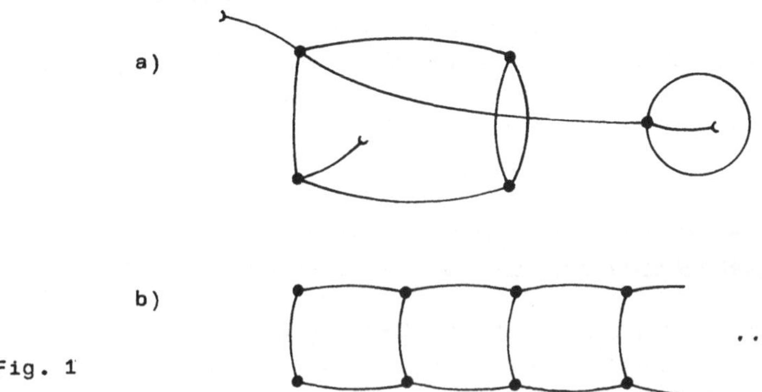

Fig. 1

A graphoid $G' = (V',H',I')$ is a _subgraphoid_ of G if $V' \subseteq V$, $H' \subseteq H$, and $I' \subseteq I$. Let $V' \cup H'$ be a set of elements of G such that $\emptyset \neq V' \subseteq V$, $H' \subseteq H$ and $\text{ver}(h') \in V'$ for any $h' \in H'$. Then the graphoid $(V',H', I \cap (V' \cup H')^2)$ is called the _subgraphoid gener-ated_ by $V' \cup H'$.

A _path_ in the graphoid $G = (V,H,I)$ is a finite or infinite sequence

$$w = (\ \ldots ,g_0,g_1, \ \ldots \ ,g_1, \ \ldots \ )$$

of elements $g_i \in V \cup H$ such that

$$(g_i,g_{i+1}) \in I,$$

and

$$\text{from} \quad g_{i-1} = g_{i+1} \quad \text{it follows that} \quad g_i \in V,$$

hence $g_{i-1} = g_{i+1} \in H$, $g_{i-2},g_{i+2} \in H$, and $g_{i-2} = g_{i+2}$ if these elements occur in w. Since I is irreflexive, we have $g_i \neq g_{i+1}$; the second above condition excludes the possibility that a path turns back within an edge. Remark that w can be infinite to the right, to the left, or both.

A path is said to be _simple_ if it contains any element at most once. We shall say that a finite path $(g_0, \ \ldots \ ,g_1)$ _connects_ its endelements g_0 and g_1. It is called _closed_ if $1 \geqslant 2$ and the sequence $(g_{1-1},g_1,g_0, \ \ldots \ ,g_1,g_0,g_1)$ is also a path, especially $(g_1,g_0) \in I$. By a _cycle_, we mean a simple closed path $(g_0, \ \ldots \ ,g_1)$.

Usually,cycles $(g_0,\ldots,g_1)$ and $(g_1,\ldots,g_1,g_0,\ldots,g_{i-1})$,where $0 \leqslant i \leqslant 1$,are considered to be equal.

Paths connecting two vertices can be uniquely determined by the sequences of traversed (directed) edges,and in simple graphoids even by the sequences of their vertices.

Two elements of a graphoid are called connected if there is a finite path connecting them. The subgraphoids generated by the equivalence classes of this relation are called the (connected) components of the given graphoid. The graphoid is said to be connented if it has only one component. This means that any two elements or,equivalently,any two vertices can be connected by a path (and even by a simple path).

Graphoids which are both finite and connected will briefly be called ficographoids,or ficographs if they possess no free half-edges.

Some further notations which are directly understandable or well-known from graph theory will be used without special explanations. For instance,two vertices can be adjacent or neighbouring, and $\deg(v) = \mathrm{card}(H(v))$ is the degree of the vertex v. A tree is a connected graph without cycles. By an open tree,we mean a connected graphoid without cycles but with free half-edges.

Throughout this book we shall only consider locally finite graphoids without isolated vertices,i.e. $\deg(v) \in \mathbb{N}^+$,especially $H \neq \emptyset$.

In usual drawings of graphoids it is allowed that lines representing different edges or half-edges intersect each other. This is excluded in embeddings of graphoids. For $n \in \{2,3\}$,by an n-dimensional embedding of the graphoid $G = (V,H,I)$, we mean a mapping P which assigns to each vertex $v \in V$ a point $P(v) \in \mathbb{R}^n$ and to each half-edge $h \in H$ a Jordan arc $P(h) \subseteq \mathbb{R}^n$ such that the following conditions hold.

(1) If $v,v' \in V$ and $v \neq v'$,then $P(v) \neq P(v')$.

(2) For all $v \in V$ and $h \in H$, $P(v)$ is not an interior point of $P(h)$, and $P(v) \in P(h)$ iff $v = \mathrm{ver}(h)$.

(3) If $h,h' \in H$ and $h \neq h'$,then $P(h)$ and $P(h')$ have no common interior point, and $(P(h) \cap P(h')) \setminus P(V) \neq \emptyset$ iff $(h,h') \in I$.

One can easily show that every finite or denumerably infinite graphoid possesses a 3-dimensional embedding. A graphoid is said to be planar if it possesses a 2-dimensional embedding. Let G_0 be the graph obtained from a graphoid G by removing all free half-edges. Obviously,G is planar iff G_0 is planar.

<u>Hints & Sources.</u> Concerning the basic concepts of graph theory
the reader is referred to textbooks,e.g. /S.Be/,/S.Ha/,/S.Or/,
/S.Sa/. Half-edges in graphs are already considered in /S.Sa/.
The concept of graphoid (called fragment,however) was introduced
in /L.He85/.

In order to enable the automata operating in graphoids to
behave in a deterministic way,we have to destroy the uniformity
in the sets $H(v)$,at least if they are considered from one of
their half-edges. This is preferably done in two ways,namely by
fixing a cyclic permutation in every $H(v)$ and by labelling the
elements of $H(v)$ with "directions" from some finite set,respect-
ively. The corresponding types of graphoids will be considered
in the following two sections.

1.2. R-graphoids

A <u>rotation</u> or cyclic permutation in a non-empty finite set S
is a bijection r of S onto itself such that $r^{card(S)}(s) = s$ and
$r^i(s) \neq r^j(s)$ for $1 \leq i < j \leq card(S)$ and any $s \in S$. Usually r is
determined by the sequence $(s,r(s), \ldots ,r^{card(S)-1}(s))$ for some
$s \in S$.

A <u>rotation system</u> on a graphoid $G = (V,H,I)$ is a mapping
$r : H \longrightarrow H$ such that,for any vertex $v \in V$,the restriction $r_{/H(v)}$
is a rotation in $H(v)$. (Remember that these sets are supposed to
be non-empty and finite.)

The quadruple $L = (V,H,I,r)$,where r is a rotation system on
(V,H,I),is called a graphoid with rotation system or briefly an
<u>R-graphoid</u>. An <u>R-graph</u> is an R-graphoid without free half-edges.

<u>Agreement.</u> In all drawings of R-graphoids let the rotation system
correspond to the clockwise orientation of the plane (within
sufficiently small neighbourhoods of the points representing the
vertices).

For instance,Fig. 2 shows an R-graphoid with the rotation
system
$$r = \{ (h_1,h_2),(h_2,h_3),(h_3,h_1),(h_4,h_5),(h_5,h_6),(h_6,h_4),(h_7,h_8),$$
$$(h_8,h_9),(h_9,h_7) \} .$$

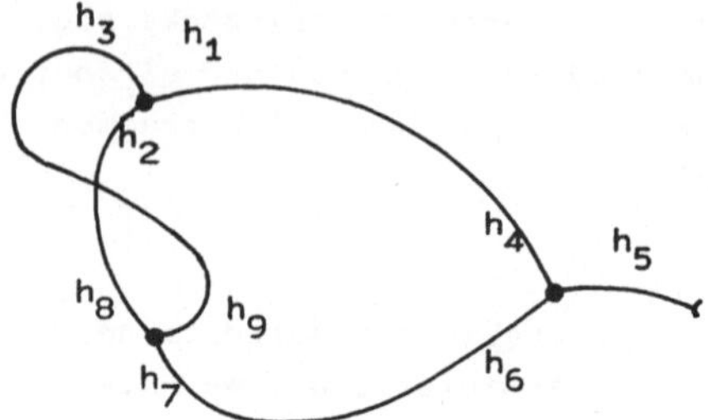

Fig. 2

Walking through labyrinths one often follows the contour of
some obstruction for a certain time. This is described by the
rule "keep the left hand on the wall". Arriving at a vertex v
through the half-edge $h \in H(v)$, one has to cross through the half-
edge $r(h)$ in the next step.

According to this idea, we introduce the concept of angle. By
an <u>angle</u> (incident to vertex v) in the R-graphoid $L = (V,H,I,r)$,
we mean a path $\alpha = (h,v,h')$, where $h \in H$, $v = \mathrm{ver}(h)$, and $h' = r(h)$.
Let $\mathrm{ang}(h)$ denote the uniquely determined angle α starting with
the half-edge h. Two angles (h_1,v_1,h_1') and (h_2,v_2,h_2') are said to
be <u>linked</u> if $(h_1',h_2) \in I$ or $(h_2',h_1) \in I$.

A <u>face</u> of the R-graphoid L is a minimal non-empty set F of
angles which is closed under the link relation, i.e., from
$(h_1,v_1,h_1') \in F$ and $(h_1',h_2) \in I$ or $(h_2',h_1) \in I$, for some angle
(h_2,v_2,h_2'), it follows that $(h_2,v_2,h_2') \in F$.

<u>Lemma 1.</u> Any angle of an R-graphoid belongs to exactly one face.
   Any face F is of one of the following <u>types</u>.
   (1) F is finite and its angles contain only bound half-edges.
   (2) F is finite and there are exactly two free half-edges
       contained in angles from F.
   (3) F is infinite and no angle from F contains a free half-
       edge.
   (4) F is infinite and there is exactly one free half-edge
       contained in an angle from F.

The first assertion easily follows. For any angle α, the intersection of all sets of angles which contain α and are closed
under the link relation forms the face containing α.

Before we prove the main assertion of the lemma, we consider
some examples of R-graphoids and their faces shown in Fig. 3.
The faces are indicated by oriented dotted lines.

18

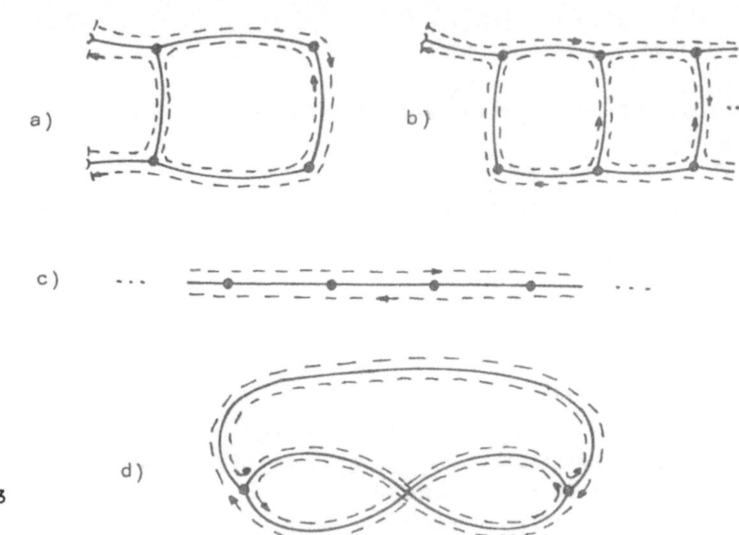

Fig. 3

The R-graphoid in Fig. 3a has one face of type (1) and two of
type (2). In Fig 3b we have infinitely many faces of type (1) and
two of type (4),in Fig. 3c we see two faces of type (3),and
Fig. 3d shows an R-graph with only one face which is of type (1).

To complete the proof of Lemma 1,let F be an arbitrary face of
$L = (V,H,I,r)$. We generate F by a suitable procedure.

Let $F_0 = \alpha_0$ for some angle $\alpha_0 \in F$.

In the (n+1)st step we assume that a set
$$F_n = \{ \alpha_{-1}, \ldots, \alpha_{-1}, \alpha_0, \alpha_1, \ldots, \alpha_m \}$$
of angles $\alpha_i = (h_i, v_i, h_i')$ is defined such that $F_n \subseteq F$, $\alpha_i \neq \alpha_j$

for $-1 \leq i < j \leq m$,and α_i is linked with α_{i+1} for $-1 \leq i < m$.

__Case i.__ Both h_{-1} and h_m' are free half-edges.
Then we stop the procedure. Obviously,F_n is a minimal non-empty
set closed under the link relation,and $F = F_n$ belongs to type
(2).

__Case ii.__ h_{-1} is a free half-edge,but h_m' is bound.
Then α_m is linked with exactly one angle α_{m+1} in L. One easily
shows that $\alpha_{m+1} \notin F_n$. We define
$$F_{n+1} = F_n \cup \{ \alpha_{m+1} \}$$
and continue with the (n+2)nd step of the procedure.

__Case iii.__ h_{-1} is bound,and h_m' is free.
Analogously,we obtain an angle α_{-1-1} linked with α_{-1}. We have
$\alpha_{-1-1} \notin F_n$, define

$$F_{n+1} = F_n \cup \{\alpha_{-1-1}\},$$
and continue with the next step of the procedure.

<u>Case iv.</u> Both h_{-1} and h_m are bound,and $(h_{-1},h_m) \in I$.
Obviously, $F = F_n$ in this case. We stop the procedure. F is of
type (1).

<u>Case v.</u> Both h_{-1} and h_m are bound,and $(h_{-1},h_m) \notin I$.
Then there are an angle α_{-1-1} linked with α_{-1} and an angle
α_{m+1} with which α_m is linked. Moreover, $\alpha_{-1-1}, \alpha_{m+1} \notin F_n$.
If $\alpha_{-1-1} \neq \alpha_{m+1}$, let
$$F_{n+1} = F_n \cup \{\alpha_{-1-1}, \alpha_{m+1}\},$$
and we continue with the (n+2)nd step of the procedure.
If $\alpha_{-1-1} = \alpha_{m+1}$, we have $F = F_n \cup \{\alpha_{m+1}\}$,and F is of type (1).
We stop the procedure.

If this procedure never halts,the face F belongs either to
type (3),namely if Case v repeatedly applies; or to type (4),
namely if Case ii or iii applies. //

Remark that finite R-graphoids have only faces of types (1) or
(2),and R-graphs have only faces of types (1) or (3).

Faces of type (1) are preferably noted as closed paths traversing each angle of the face exactly once and following the link
relation. So we shall write $F = (h_0,v_0,h_0',h_1, \ldots ,h_l,v_l,h_l')$
instead of $F = \{(h_i,v_i,h_i'): 0 \leq i \leq l\}$ if $(h_i',h_{i+1}) \in I$ for $0 \leq i < l$
and $(h_l',h_0) \in I$. Faces of the other types can be noted by finite
or infinite paths in a corresponding manner.

A path $w = (\ldots ,g_0,g_1, \ldots ,g_l, \ldots)$ is said to be <u>face-
following</u> if for any three consecutive elements g_i,g_{i+1},g_{i+2} of w,
where g_{i+1} is a vertex of the graphoid,it follows that the triple
(g_i,g_{i+1},g_{i+2}) is an angle.

For a vertex v and a half-edge h in an R-graphoid $L = (V,H,I,r)$,
let the <u>order</u> , ord(v) and ord(h) , denote the number of faces containing angles in which v and h occur,respectively. More precisely,

$$\text{ord}(v) = \text{card} \{F: F \text{ is a face of L and contains an angle incident to } v\},$$

$$\text{ord}(h) = \begin{cases} 1 & \text{if } \text{ang}(h)=(h,\text{ver}(h),r(h)) \text{ and} \\ & \text{ang}(r^{-1}(h))=(r^{-1}(h),\text{ver}(h),h) \text{ belong} \\ & \text{to the same face of L }, \\ 2 & \text{if the above angles belong to different} \\ & \text{faces }. \end{cases}$$

If the half-edges h_1 and h_2 are incident,they have equal order.
For an undirected or directed edge e consisting of half-edges h_1
and h_2 , i.e. $e = \{h_1,h_2\}$ or $e = (h_1,h_2)$,let the order be defined
by
$$\text{ord}(e) = \text{ord}(h_1) \quad (= \text{ord}(h_2)).$$

Fig. 4 shows the orders of some vertices and edges in an
R-graph.

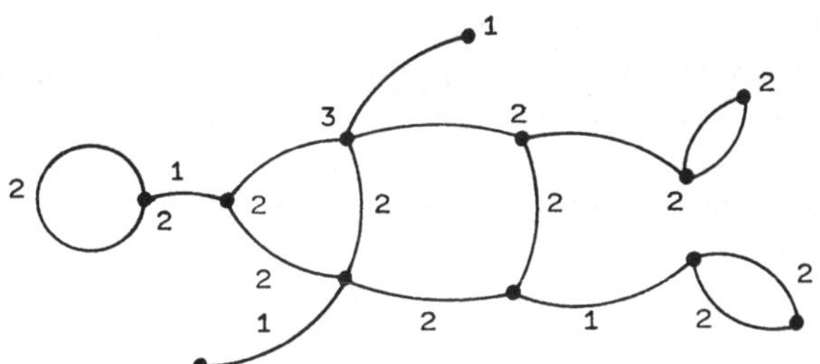

Fig. 4

<u>Lemma 2.</u> Let a connected R-graphoid L have at least two faces.
   Then any face of L contains an angle with a half-edge of
   order 2.

We consider an arbitrary face F, a face F' different from F, and
half-edges h and h' such that $ang(h) \in F$ and $ang(h') \in F'$. There is
a path $w = (h_0, v_0, h_0', \ldots, h_1, v_1)$ in L with $h_0 = h$, $v_0 = ver(h)$ and
$v_1 = ver(h')$. It follows that $ord(v_i) \geq 2$ for some $i \in \{0, \ldots, 1\}$.
Let i be the minimal index of this kind. If $ang(r^j(h_i))$ is the
first angle which does not belong to F in the sequence
$(ang(h_i), ang(r(h_i)), \ldots, ang(r^{deg(v_i)-1}(h_i)))$, then $0 < j$
and $ang(r^{j-1}(h_i)) \in F$, but $ang(r^j(h_i)) \notin F$. Hence we have
$ord(r^j(h_i)) = 2$. //

In walking through a labyrinth, it may be useful to know which
of the angles incident to some vertex v and considered from some
half-edge $h \in H(v)$ belong to the same faces. Accordingly, for
$0 \leq i, j < deg(v)$, we define

$$i \sim_h j \quad iff \quad ang(r^i(h)) \text{ and } ang(r^j(h)) \text{ belong to}$$
$$\text{the same face.}$$

This is an equivalence relation in the set $\{0, 1, \ldots, deg(v)-1\}$
and will be called the (<u>order</u>) <u>characteristic</u> of the half-edge h.
By $cha(h)$ we shall denote the corresponding factorization, i.e.
the system of equivalence classes,

$$cha(h) = \{\{j: 0 \leq j < deg(v) \text{ and } j \sim_h i\}: 0 \leq i < deg(v)\}.$$

The orders of v and h can be easily computed from $cha(h)$,
namely $ord(v) = card(cha(h))$, and $ord(h) = 1$ iff $0 \sim_h deg(v)-1$.

A <u>plane</u> <u>embedding</u> of an R-graphoid (V, H, I, r) is a 2-dimen-
sional embedding P of the underlying graphoid (V, H, I) with the
additional property that the rotation system r corresponds to the
clockwise orientation of the plane. The latter means that, for any

$v \in V$ and $h' \in H(v)$, $P(r(h'))$ is the direct successor of $P(h')$ in
the set of Jordan arcs $\{P(h): h \in H(v)\}$ with respect to the clockwise rotation in the point $P(v)$.

An R-graphoid is said to be <u>plane</u> if it possesses a plane embedding. For example, the R-graph from Fig. 3d is not plane, although the underlying graph is planar. Obviously, an R-graphoid is
plane iff the R-graph obtained from it by removing all free half-
edges is plane.

To reduce the possible scruples concerning topological problems
connected with plane embeddings, we refer to the results on straight
line representations of planar graphs. So K. Wagner /S.Wa/,
I. Fary /S.Fa/ and S. Stein /S.Ste/ independently showed that
every finite planar simple graph is straight line representable,
i.e., it possesses an embedding in which the edges are represented
by straight line segments. This result can be immediately transferred to finite plane simple R-graphs. Using Zorn's lemma,
C. Thomassen /S.Th/ even proved that any (infinite) planar simple
graph possesses a straight line representation. Also this proof
can be modified to denumerably infinite plane simple R-graphs.
On the other hand, it is shown that any planar graph has only
countably many vertices of degree $\geqslant 3$.

Now let P be a plane embedding of an R-ficograph $L = (V,H,I,r)$.
From Jordan's curve theorem it follows that the complement of the
point set $S = P(V) \cup \bigcup P(H) = \bigcup P(H)$ consists of finitely many
simply connected open components, the so-called <u>regions</u> of L with
respect to P, which are separated by S. (All these notations refer
to the natural topology of the Euclidean plane.)

By Euler's theorem on polyhedra, there are exactly
$\mathrm{card}(E) - \mathrm{card}(V) + 2$ regions, where E is the set of all edges of L.
Moreover, there is a one-to-one correspondence between the regions
and the faces of L which all are of type (1). Indeed, the faces
considered as closed paths give the oriented boundaries of the
regions in such a way that any region touches its boundary on the
left-hand side. Hence the number of faces equals $\mathrm{card}(E) - \mathrm{card}(V)$
$+ 2$, too.

This equation even allows a nice combinatorial chacterization
of the plane R-ficographs. For an arbitrary R-ficograph
$L = (V,H,I,r)$ with exactly n_e edges and n_F faces, the so-called
<u>Euler-Poincaré</u> <u>characteristic</u> is defined by

$$\mathrm{epc}(L) = \mathrm{card}(V) - n_e + n_F .$$

<u>Proposition 1.</u> An R-ficograph L is plane iff $\mathrm{epc}(L) = 2$.

The necessity of the equation is given by Euler's formula. The sufficiency can be proved by induction on n_F.

If $n_F = 1$ and epc(L) = 2,we have card(V) = n_e + 1. Then L is a finite R-tree. This can be proved by induction on card(V),and in the same way it follows that any finite R-tree is plane.

Now let the assertion hold for all R-ficographs with $k \geqslant 1$ faces,and let n_F = k+1. By Lemma 2,L contains an edge e_0 = $\{h_1,h_2\}$ of order 2. For the R-ficograph L' obtained from L by removing e_0, we also have epc(L') = 2. L' has only k faces,where one of these, say F',has been obtained by "fusion" of the two faces containing e_0 in L. Therefore,L' possesses a plane embedding P'. One can easily see that P' can be completed to a plane embedding P of L by adding a Jordan arc $P(e_0)$ = $P(h_1) \cup P(h_2)$ which connects $P'(\text{ver}(h_1))$ with $P'(\text{ver}(h_2))$ such that all its interior points are contained in the region corresponding to the face F' with respect to P'. //

In any plane embedding of an R-ficograph,there is exactly one unbounded region called the <u>exterior</u> one. The other,bounded re- gions are called <u>interior regions</u>. The corresponding faces are also said to be <u>exterior</u> and <u>interior</u> faces,respectively.

In the general case of a plane embedding P of a connected R-graphoid L,the complement of S = $\cup P(H)$ may consist of infi- nitely many components. Then a face is said to be an <u>exterior</u> one if the component touching it on the left-hand side is unbounded. Otherwise it is said to be an <u>interior</u> face. For instance,the embedding shown in Fig. 3b has two exterior faces of type (4) (remark that the corresponding components of S coincide) and infinitely many interior faces of type (1). Remark that interior faces could also be infinite.

Naturally,by an <u>R-isomorphism</u> between R-graphoids L = (V,H,I,r) and L' = (V',H',I',r') ,we understand an isomorphism φ between the underlying graphoids which preserves the rotation; more precisely,

$$\varphi(r(h)) = r'(\varphi(h))$$

for any half-edge h of L.

Of course,the property to be plane is invariant under R-iso- morphisms. Note,however,that the property of faces to be interior or exterior ones is not invariant under R-isomorphisms. Indeed,to any face F of a plane R-ficograph,there is a plane embedding with respect to which F is the exterior face. This can be proved via stereographic projection.

Another,more direct proof is as follows. We start with an arbitrary plane embedding of the R-ficograph. If F is an interior

face,there is a Jordan arc J connecting a point of the F-region
with the exterior region such that no embedding of a vertex
belongs to J and any embedding of an edge is at most once crossed.
Now we cut off all embeddings of edges which cross J and modify
the connections in such a way that F becomes the exterior face of
the new embedding,see Fig. 5.

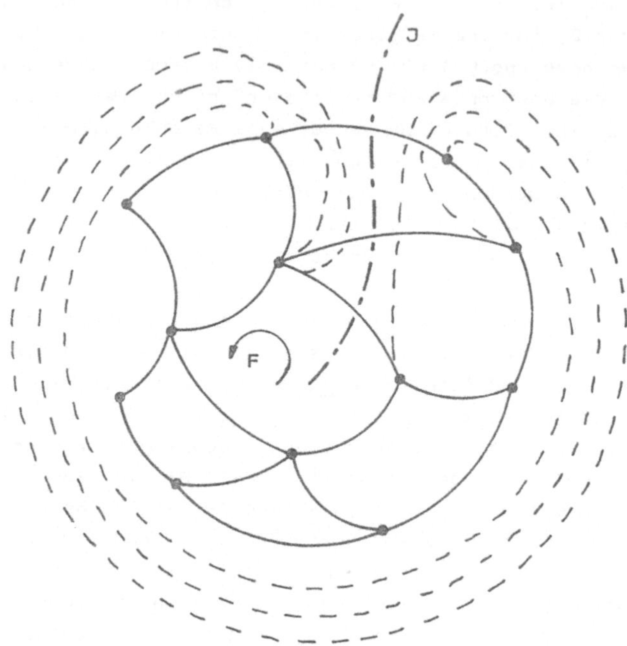

Fig. 5

An edge $e = h_1,h_2$ is called a <u>bridge</u> of an R-graphoid
$L = (V,H,I,r)$ (or of the underlying graphoid) if the graphoid
$L' = (V,H,I \setminus \{(h_1,h_2),(h_2,h_1)\})$ obtained from L by cutting e has
more connected components than L. In plane infinite R-graphs or
in plane R-graphoids with free half-edges,a bridge can have the
order 2,cf. Fig. 3c. In the non-plane R-graph in Fig. 3d,any edge
has order 1,but there is no bridge.

<u>Lemma 3.</u> An edge e of a finite plane R-graph is a bridge iff
     $\operatorname{ord}(e) = 1$.

Let $F = (h_0,v_0,h_0',h_1, \ldots ,h_l,v_l,h_l')$ be a face containing the
edge $e = \{h_0,h_l'\}$.
If $\operatorname{ord}(e) = 2$,we successively replace all subpaths of the form
$(v_i,h_i',h_{i+1}, \ldots ,h_{i+j-1}',h_{i+j},v_{i+j})$, where $v_i = v_{i+j}$, by (v_i).

24

In this way we obtain from F a closed simple path,i.e. a cycle,
containing the edge e. Hence e is not a bridge.

If $\mathrm{ord}(e)=1$,there is a minimal index $i<n$ such that $h_0=h_i'$.
Let J be a Jordan arc running to the left of the embedding of the
path $(v_0,h_0',h_1,\ldots,h_i,v_i=v_0)$ in a sufficiently small distance.
It can be completed to a (closed) Jordan curve $\bar{J}$ which crosses no
embedding of an element of the graph except that of edge e,see
Fig. 6. Therefore,any path of the graph which connects the half-
edges h_0 and h_i' of e must contain (h_0,h_0') as a subpath,i.e.,e is
a bridge. //

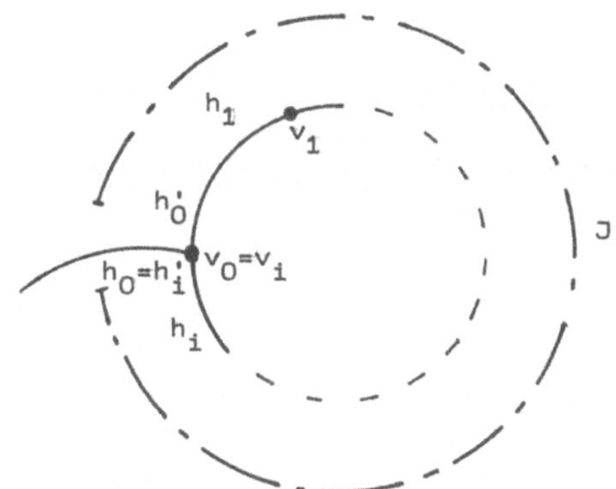

Fig. 6

One easily shows that any R-tree is plane and all of its edges
are bridges. Hence they have order 1 if the tree is finite. Note
that the R-graph from Fig. 3d is not a tree,although all edges
have order 1.

<u>Hints & Sources.</u> Rotation systems of graphs have already widely
been used in topological graph theory,cf. /S.St/. Generalizing
Proposition 1,the genus of a surface on which a given R-ficograph
is embedded can be characterized by the Euler-Poincaré characteristic. The first mention of rotation systems within labyrinth
theory is by F. Hoffmann and K. Kriegel /L.HoKr/.

1.3. C-graphoids

In the second basic type of labyrinths,the uniformity in the
sets $H(v)$ of graphoids is destroyed by compass systems.

Let D be some finite non-empty set. By a __compass system__ with
the direction set D on the graphoid (V,H,I) , we mean a mapping
$c: H \longrightarrow D$ such that the restriction $c_{/H(v)}$ is injective,for
any vertex $v \in V$. It follows that $\deg(v) \leq \operatorname{card}(D)$. The quadruple
$L = (V,H,I,c)$ is said to be a __C-graphoid__ then.

By fixing some rotation r_D in D , any __compass system__ c __deter-
mines__ a __rotation system__ r_c in a natural way. One has only to take
into consideration that $\deg(v) < \operatorname{card}(D)$,possibly. For $h \in H$ let
$k = \min\{i: i \geq 1 \text{ and } r_D^i(c(h)) \in c(H(\operatorname{ver}(h))) \}$, and

$$r_c(h) = c_{/H(\operatorname{ver}(h))}^{-1} \cdot r_D^k \cdot c(h) .$$

In this sense the C-graphoids will be considered to be special
R-graphoids,and all concepts and notations already introduced for
these will be applied to those,too.

A __C-isomorphism__ between C-graphoids L and $L' = (V',H',I',c')$
is an isomorphism φ between the underlying graphoids such that
$c'(\varphi(h)) = c(h)$ for any half-edge $h \in H$. Obviously,C-isomorphic
C-graphoids are also R-isomorphic.

The following two direction sets are of special importance to
us,

$$\mathbf{D}_2 = \{\text{north,east,south,west} \},$$
$$\mathbf{D}_3 = \mathbf{D}_2 \cup \{\text{up,down} \} .$$

By a __3D graphoid__ (that means, 3-dimensional C-graphoid),we
understand a C-graphoid $L = (V,H,I,c)$ with the direction set $\mathbf{D}_3$
and possessing a 3-dimensional embedding P such that

- the representations $P(h) \cup P(h')$ of all edges $\{h,h'\}$ and
  the representations $P(h)$ of all free half-edges h are
  straight line segments parallel or perpendicular to the coordinate axes, and
- for any $h \in H$, from $P(\operatorname{ver}(h)) = (x,y,z) \in \mathbf{R}^3$ it follows that

$$c(h) = \begin{cases} \text{north} & \text{if } P(h) = \{(x,y+t,z): t \in [0,t'] \} \\ \text{south} & \text{if } P(h) = \{(x,y-t,z): t \in [0,t'] \} \\ \text{east} & \text{if } P(h) = \{(x+t,y,z): t \in [0,t'] \} \\ \text{west} & \text{if } P(h) = \{(x-t,y,z): t \in [0,t'] \} \\ \text{up} & \text{if } P(h) = \{(x,y,z+t): t \in [0,t'] \} \\ \text{down} & \text{if } P(h) = \{(x,y,z-t): t \in [0,t'] \} \end{cases} \quad \text{for some } t'>0.$$

Intuitively,the latter condition means that $c(h)$ is the real

direction in which the Jordan arc P(h) leaves the point P(ver(h)).
An embedding P satisfying the two above conditions is called a
<u>3D embedding</u>.

If {h,h'} is an edge of a 3D graphoid L,the directions c(h)
and c(h') must be <u>opposite</u>,i.e. $c(h) = \overline{c(h')}$, where $\overline{\text{north}}$ = south,
$\overline{\text{south}}$ = north , $\overline{\text{east}}$ = west , $\overline{\text{west}}$ = east , $\overline{\text{up}}$ = down , $\overline{\text{down}}$ = up.

Often 3D graphoids will be represented or even defined by draw-
ing (sufficiently large finite parts of their images under) 3D
embeddings. Fig. 7 shows some examples and the corresponding
coordinate axes.

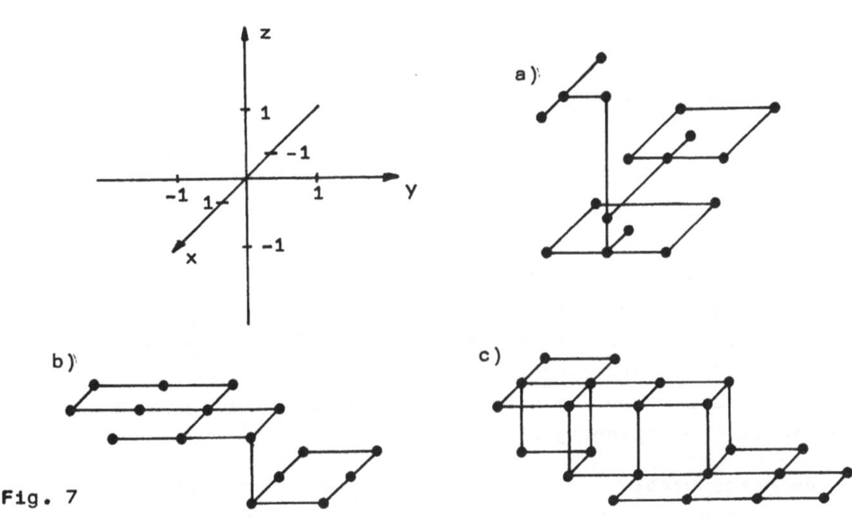

Fig. 7

Obviously,3D graphoids have simple underlying graphoids. Therefore,paths connecting vertices are simply noted as sequences of
vertices $(v_0,v_1, \ldots ,v_1)$, and closed paths or faces of type (1)
are noted in the form $(v_0,v_1, \ldots ,v_{1-1};v_1=v_0)$. The meaning is
straightforward.

A 3D embedding P and the corresponding 3D graphoid L are said
to be <u>normed</u> if the Jordan arcs P(h) are straight line segments of
length $\frac{1}{2}$, for all half-edges h. It follows that all embeddings of
edges have length 1,i.e.,for any two adjacent vertices v and v'
we have $|P(v) - P(v')| = 1.$

If,conversely,from $|P(v) - P(v')| = 1$ it follows that v and v'
are adjacent,for any two vertices v and v' of L,then the graphoid
and the embedding P are called <u>strongly normed</u>. To illustrate these

notations,we remark that the 3D ficograph in Fig. 7a is not normed,
that of Fig. 7b is normed but not strongly normed,and that of
Fig. 7c is strongly normed.

By a <u>2D graphoid</u> we mean a 3D graphoid possessing a 3D embedding
P in a horizontal plane,without loss of generality,in the
(x,y)-plane , i.e. $\cup P(H) \subseteq \mathbb{R} \times \mathbb{R} \times \{0\}$,where H is the set of all
half-edges. Sometimes we omit the third coordinates $z = 0$ of the
embedding points in this case. The corresponding embedding is
called a <u>2D embedding</u>. Fig. 8a shows a normed,not strongly normed
2D ficograph,and Fig. 8b a strongly normed 2D ficograph.

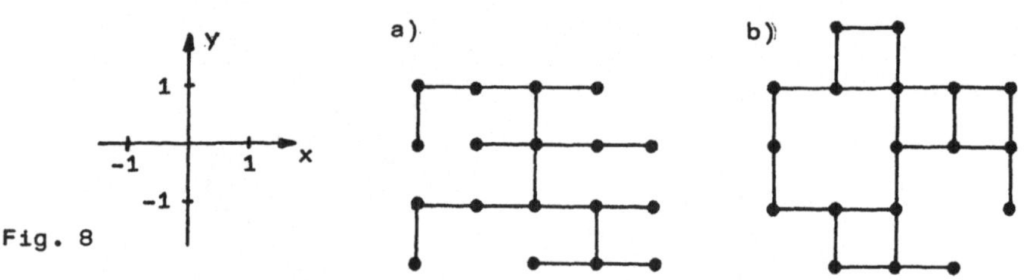

Fig. 8

In the direction set $\mathbb{D}_2$,we fix the rotation
$$r_{\mathbb{D}_2} = (north,east,south,west)$$
which corresponds to the clockwise orientation of the plane. Then
the 2D graphoids are special plane R-graphoids,where the rotation
system is defined by the given compass system via $r_{\mathbb{D}_2}$, as explained at the beginning of this section.

The means usable in 2D graphoids considerably exceed those
possible in general plane R-graphoids. This improvement is essentially based on the use of the rotation index of paths. We first
consider some fundamental facts which,like Jordan's curve theorem,
hold in a more general framework than given by embedded graphoids.

For any two points in the plane, $p,p' \in \mathbb{R}^2$, let $\overline{p\,p'}$ denote the
(undirected) straight line segment connecting p and p' ,and $\overrightarrow{p\,p'}$
denote the directed segment from p to p'. If $\overline{p\,p'}$ is parallel or
perpendicular to both axes, $dir(\overrightarrow{p\,p'})$ denotes the direction of
$\overrightarrow{p\,p'}$,i.e.

$$dir(\overrightarrow{p\,p'}) = \begin{cases} north & \text{if } p'-p \in \{0\} \times \mathbb{R}^+ , \\ south & \text{if } p'-p \in \{0\} \times \mathbb{R}^- , \\ east & \text{if } p'-p \in \mathbb{R}^+ \times \{0\} , \\ west & \text{if } p'-p \in \mathbb{R}^- \times \{0\} . \end{cases}$$

By a <u>rectilinear polygonal path</u>,briefly RPP,we mean a sequence
$$w = (p_0,p_1, \ldots ,p_1) , \quad 1 \geqslant 1 ,$$

of points in the plane, $p_i \in \mathbf{R}^2$,such that

$$p_i \neq p_{i+1} \quad \text{for } 0 \leq i < l \text{ , and}$$

the segments $\overline{p_i p_{i+1}}$ are parallel or perpendicular to the coordinate axes.

The <u>rotation</u> <u>index</u> $rin(w)$ counts the right-angled rotations in the course of running through w. More precisely,

$$rin(p_0,p_1) = 0 \text{ , and}$$

$$rin(p_0,p_1,\ldots,p_l) = \sum_{i=0}^{l-2} rin(p_i,p_{i+1},p_{i+2})$$

for $l \geq 2$, where

$$rin(p_i,p_{i+1},p_{i+2}) = \begin{cases} 1 & \text{if } dir(\overrightarrow{p_{i+1}p_{i+2}}) = r_{\mathbf{D}_2}^{-1} \cdot dir(\overrightarrow{p_i p_{i+1}}), \\ 0 & \text{if } dir(\overrightarrow{p_{i+1}p_{i+2}}) = dir(\overrightarrow{p_i p_{i+1}}), \\ -1 & \text{if } dir(\overrightarrow{p_{i+1}p_{i+2}}) = r_{\mathbf{D}_2} \cdot dir(\overrightarrow{p_i p_{i+1}}), \\ -2 & \text{if } dir(\overrightarrow{p_{i+1}p_{i+2}}) = r_{\mathbf{D}_2}^2 \cdot dir(\overrightarrow{p_i p_{i+1}}). \end{cases}$$

For instance,for the point set drawn in Fig. 9 we have $rin(a,d,e) = -1$, $rin(c,g,h,e,d) = -3$, $rin(c,b,f,b) = -3$, $rin(e,h,g,c,b) = 1$.

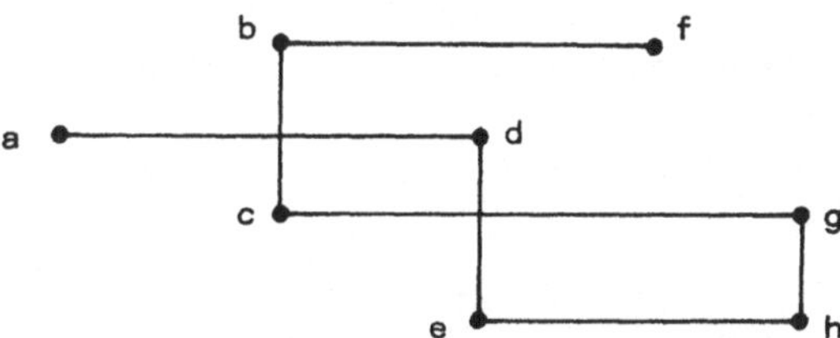

Fig. 9

The assertions of the following lemma either trivially hold or can easily be proved by induction on l.

<u>Lemma 4.</u> For any RPP $w = (p_0,p_1,\ldots,p_l)$,

$$rin(w) = rin(p_0,\ldots,p_m,p_{m+1}) + rin(p_m,p_{m+1},\ldots,p_l) \text{ if } 0 \leq m < l,$$

$$rin(w) = -rin(p_l,p_{l-1},\ldots,p_0) \text{ , and}$$

$$rin(w) \equiv -k \mod 4 \text{ if } dir(\overrightarrow{p_{l-1}p_l}) = r_{\mathbf{D}_2}^k \cdot dir(\overrightarrow{p_0 p_1}),$$

$$\text{for } k \in \{0,1,2,3\} \text{ . } //$$

An RPP $w = (p_0,p_1,\ldots,p_l)$ is said to be <u>simple</u> if, for $0 \leq i < j < l$, from $\overline{p_i p_{i+1}} \cap \overline{p_j p_{j+1}} \neq \emptyset$ it follows that $j = i+1$ and $\overline{p_i p_{i+1}} \cap \overline{p_j p_{j+1}} = \{p_j\}$.

Especially in this case we have $p_i \neq p_j$, for $0 \leq i < j < l$, and any two different segments $\overline{p_i p_{i+1}}$ and $\overline{p_j p_{j+1}}$ cannot cross each other. The concatenation of the directed straight line segments $\overrightarrow{p_0 p_1}, \overrightarrow{p_1 p_2}, \ldots, \overrightarrow{p_{l-1} p_l}$ forms an oriented Jordan arc $\mathbf{J}_w$ in the plane. For instance,the RPP (d,e,h,g,c) in Fig. 9 is not simple.

An RPP $w = (p_0,p_1, \ldots ,p_l))$ is called <u>closed</u> if $p_l = p_0$. Then let the <u>closed rotation index</u> be defined by
$$\overline{rin}(w) = rin(w) + rin(p_{l-1},p_l,p_1).$$
w is said to be <u>simply closed</u> if,moreover,for $0 \leqslant i < j < l$, from $\overline{p_i p_{i+1}} \cap \overline{p_j p_{j+1}} \neq \emptyset$ it follows that
$$\text{either} \quad j = i+1 \quad \text{and} \quad \overline{p_i p_{i+1}} \cap \overline{p_j p_{j+1}} = \{p_j\},$$
$$\text{or} \quad j = l-1 , i = 0 \quad \text{and} \quad \overline{p_i p_{i+1}} \cap \overline{p_j p_{j+1}} = \{p_0\} = \{p_l\}.$$

For any simply closed RPP w, J_w is a (closed) Jordan curve. It divides the rest of the plane into a bounded region and an unbounded region. One of these regions touches J_w (or w , as we shall simply say) from the left-hand side,the other touches w from the right-hand side.

<u>Lemma 5.</u> For any simply closed RPP w ,
$$\overline{rin}(w) = \begin{cases} 4 & \text{if the region touching w from the} \\ & \text{left-hand side is bounded} , \\ -4 & \text{if this region is unbounded.} \end{cases}$$

This is a discrete version of the Riemann-Hopf theorem of turning tangents known from differential geometry.

To prove the lemma,we consider a simply closed RPP $w = (p_0,p_1,\ldots,p_l)$. For $j > l$ let $p_j = p_i$ if $0 \leqslant i < l$ and $i \equiv j$ modulo l.

It holds $rin(p_i,p_{i+1},p_{i+2}) \neq -2$, since $\overline{p_i p_{i+1}}$ and $\overline{p_{i+1} p_{i+2}}$ would have common interior points otherwise. Hence $l \geqslant 4$.

If $rin(p_i,p_{i+1},p_{i+2}) = 0$ for some $i \in \{0,1, \ldots ,l-1\}$ and the assertion holds for the simply closed RPP $(p_0,\ldots,p_{i-1},p_{i+1},\ldots \ldots,p_l)$, then it follows for w,too. Therefore,it suffices to prove the lemma for paths w satisfying
$$rin(p_i,p_{i+1},p_{i+2}) \in \{1,-1\} \quad \text{for all} \quad i \in \mathbb{N}.$$
Now we assume that w is of this kind and
$$rin(p_i,p_{i+1},p_{i+2}) = - rin(p_{i+1},p_{i+2},p_{i+3}).$$
Without loss of generality let $0 \leqslant i \leqslant l-3$, $rin(p_i,p_{i+1},p_{i+2}) = 1$, and $dir(\overrightarrow{p_i p_{i+1}}) = east$. Fig. 10 illustrates this situation.

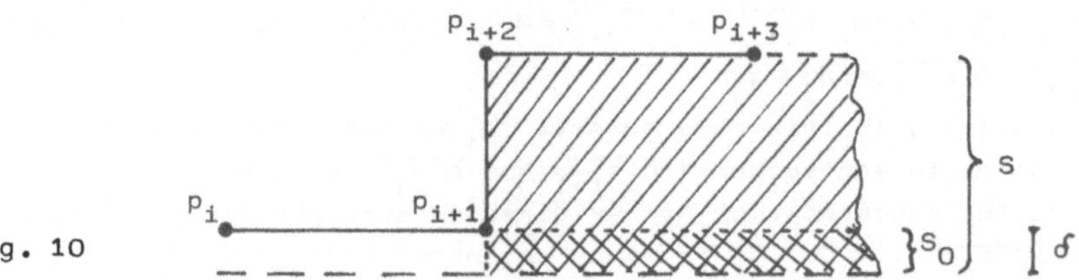

Fig. 10

Assume $p_j = (x_j, y_j)$ for $j \in \mathbb{N}$. There is a $\delta > 0$ such that
$$\delta < \tfrac{1}{2} \cdot \min\{\, |y_j - y_k| : 0 \leqslant j, k < 1 \text{ and } y_j \neq y_k \,\}.$$
We define a bijective transformation τ of the half-strip
$$S = \{\, (x,y): x_{i+1} \leqslant x, \ y_{i+1} - \delta \leqslant y \leqslant y_{i+2} \,\}$$
onto the half-strip
$$S_0 = \{\, (x,y): x_{i+1} \leqslant x, \ y_{i+1} - \delta \leqslant y \leqslant y_{i+1} \,\}$$
by
$$\tau\,(x, \ y_{i+1} - \delta + a\,) = (\,x, \ y_{i+1} - \delta + a \cdot \frac{\delta}{y_{i+2} - y_{i+1} + \delta}\,),$$
for $x_{i+1} \leqslant x$ and $0 \leqslant a \leqslant y_{i+2} - y_{i+1} + \delta$, cf. Fig. 10. Especially,
$\tau(\,p_{i+2}\,) = p_{i+1}$ and $\tau(p_{i+3}) = (x_{i+3}, y_{i+1})$.

For $0 \leqslant j \leqslant 1$ and $j \neq i+2$, let
$$\hat{p}_j = \begin{cases} p_j & \text{if } \ p_j \notin S, \\ \tau(p_j) & \text{if } \ p_j \in S. \end{cases}$$
Then $\hat{w} = (\,\hat{p}_0, \ \ldots, \hat{p}_i, \hat{p}_{i+1}, \hat{p}_{i+3}, \ \ldots, \hat{p}_1\,)$ is a simply closed RPP,
too. Furthermore, $\overline{\mathrm{rin}(\hat{w})} = \overline{\mathrm{rin}(w)}$, and the region of w touching
it from the left-hand side is bounded iff the region of $\hat{w}$ touching
this RPP from the left-hand side is bounded.

This process of removing subpaths of length 4 with the rotation
index 0 can be repeated as long as such subpaths exist. Therefore
it is sufficient to prove the lemma for all simply closed RPP
$w = (p_0, p_1, \ \ldots, p_1)$ satisfying
$$\mathrm{rin}(p_i, p_{i+1}, p_{i+2}) = z \quad \text{for all} \ \ i \in \mathbb{N},$$
with some fixed $z \in \{1, -1\}$.

In this case, the infinite sequence $(p_0, p_1, \ \ldots, p_1, \ \ldots\,)$
describes a spiral. By Lemma 4, $1 \equiv 0$ modulo 4.

Let $d_i = |p_i - p_{i-1}|$ for $i \in \mathbb{N}^+$.

If $d_{i+2} = d_i$ for all $i \in \mathbb{N}^+$, then $1 = 4$, since w is simply closed.
But for $1 = 4$ the assertion of Lemma 5 holds trivially.

Assume $d_{i+2} \neq d_i$ for some $i \geqslant 1$. Since $d_{j+1} = d_j$, there is an
index j such that $d_{j+2} < d_j$. It follows $d_{j+1} > d_{j+3} > \ldots > d_{j+1+1} = d_{j+1}$, see Fig. 11. This contradiction completes the proof. //

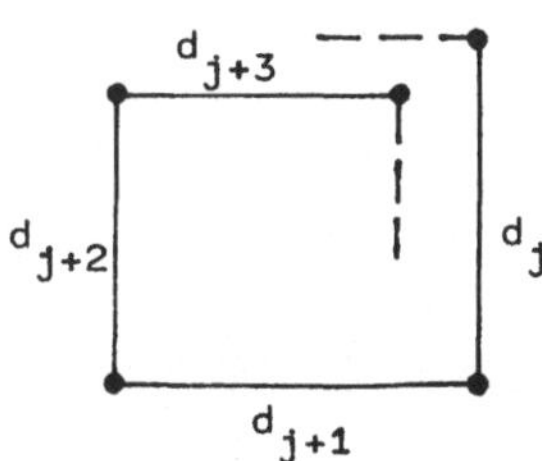

Fig. 11

By a <u>piece</u> (of path) of a graphoid (V,H,I), we mean a path of the form (h,v,h'), where $v \in V$ and $h,h' \in H(v)$. For instance, any angle of an R-graphoid is a piece of the underlying graphoid.

In C-graphoids $L = (V,H,I,c)$ with the direction set D_2, the rotation index of a piece (h,v,h') is defined by

$$rin(h,v,h') = \begin{cases} 1 & \text{if} & c(h') = r_{D_2}^{-1} \cdot \overline{c(h)}, \\ 0 & \text{if} & c(h') = \overline{c(h)}, \\ -1 & \text{if} & c(h') = r_{D_2} \cdot \overline{c(h)}, \\ -2 & \text{if} & c(h') = r_{D_2}^{2} \cdot \overline{c(h)}. \end{cases}$$

Remember that the mapping $^{-}$ gives the opposite directions.

The <u>rotation index</u> of a path $w = (g_0, g_1, \dots, g_1)$ is the sum of the rotation indices of all pieces traversed by w,

$$rin(w) = \sum_{((g_i,g_{i+1},g_{i+2}) \text{ is a piece })} rin(g_i,g_{i+1},g_{i+2}).$$

If w is closed, let

$$\overline{rin}(w) = rin(g_1,g_0,g_1, \dots, g_{1-1},g_1,g_0), \text{ i.e.}$$

$$\overline{rin}(w) = \begin{cases} rin(w) & \text{if } g_0,g_1 \in H, \\ rin(w) + rin(g_{1-1},g_1,g_0) & \text{if } g_1 \in V, g_0 \in H, \\ rin(w) + rin(g_1,g_0,g_1) & \text{if } g_0 \in V, g_1 \in H. \end{cases}$$

Correspondingly, the closed rotation index of a face of type (1),

$$F = (h_0,v_0,h_0',h_1,v_1,h_1', \dots, h_1,v_1,h_1'),$$

is the sum of the rotation indices of all angles contained in F,

$$\overline{rin}(F) = \sum_{i=0}^{1} rin(h_i,v_i,h_i').$$

Let P be a 2D embedding of L and w a vertex-connecting path written as sequence of vertices, $w = (v_0,v_1, \dots, v_1)$. Then the sequence $w_P = (P(v_0),P(v_1), \dots, P(v_1))$ is an RPP and

$$rin(w) = rin(w_P),$$

as one easily sees. Therefore, Lemma 4 can be applied to vertex-connecting paths in 2D graphoids.

Analogously, if w is closed, say $w = (v_0,v_1, \dots, v_{1-1}; v_1 = v_0)$, then

$$\overline{rin}(w) = \overline{rin}(w_P).$$

<u>Lemma 6.</u> Let F be a face of type (1) of a 2D graphoid. Then

$$\overline{rin}(F) = \begin{cases} 4 & \text{if F is an interior face,} \\ -4 & \text{if F is an exterior face.} \end{cases}$$

This follows from Lemma 5 by considering a single closed RPP w_F running to the left of F in a sufficiently small distance such that $\overline{rin}(w_F) = \overline{rin}(F)$. We omit the formal details, Fig. 12 shows an example. //

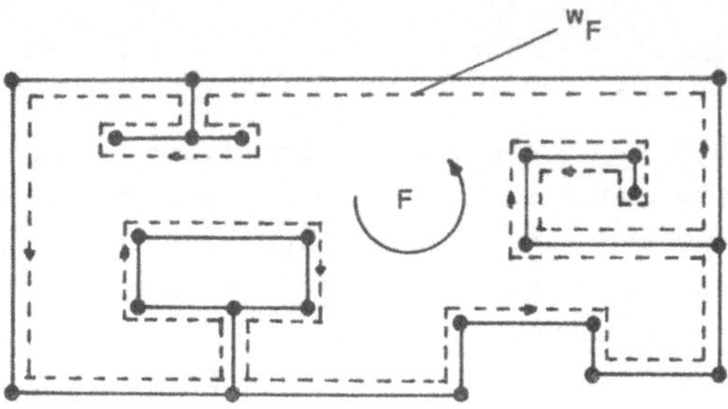

Fig. 12

Lemma 6 shows that,in 2D graphoids,the property of faces of
type (1) to be interior or exterior ones is invariant under C-iso-
morphisms. Remember that even in plane R-ficographs the interior
or exterior faces are not invariant under R-isomorphisms.

As a counterpart to Proposition 1,now we give a combinatorial
characterization of the 2D ficographs.

<u>Proposition 2.</u> Let $L = (V,H,I,c)$ be a C-ficograph with the direc-
tion set $\mathbf{D}_2$. L possesses a 2D embedding,i.e.,it is a 2D
ficograph , iff the following three conditions hold.

   i) $c(h_1) = \overline{c(h_2)}$ for all edges $\{h_1,h_2\}$ of L ;
   ii)   there is exactly one face F_0 of L such that
        $\overline{rin}(F_0) = -4$;
   iii) $\overline{rin}(F) = 4$ for all other faces $F \neq F_0$ of L.

The necessity of Condition i) is obvious , for ii) and iii) it
follows from Lemma 6.

The sufficiency of i) – iii) will be shown by induction on the
number n_F of faces of L.

From $n_F = 1$ and i) – iii) it follows that L is a C-tree and 2D
embeddable. This can be proved by induction on $card(V)$. Since $H \neq \emptyset$,
the induction starts with $card(V) = 2$. From $n_F = 1$ it follows that
L possesses only one edge,and the assertion obviously holds.

Now let the assertion hold for all C-ficographs L' with $<card(V)$
vertices,and let L have only one face F_0 and satisfy i) – iii). If
there is a vertex of degree 1 in L,then let L' be obtained from L
by removing this vertex together with the incident edge. L' has
only one face F_0' , and $\overline{rin}(F_0') = \overline{rin}(F) = -4$. It follows that L' is
a 2D embeddable C-tree,hence this holds true for L,too.

If there is no vertex of degree 1 in L,we have

$$\overline{rin}(F_0) = \sum_{(h \in H)} rin(\ ang(h)\ )$$

$$= \sum_{(h \in H,\ deg(ver(h)) = 2)} rin(\ ang(h)\ )$$

$$+ \sum_{(h \in H,\ deg(ver(h)) > 2)} rin(\ ang(h)\ ) \geqslant 0,$$

since $\displaystyle\sum_{(h \in H(v))} \mathrm{rin}(\ \mathrm{ang}(h)\ )$ $\begin{cases} = 0 & \text{if } \deg(v) = 2, \\ \geq 2 & \text{if } \deg(v) > 2. \end{cases}$

But $\overline{\mathrm{rin}}(F_0) \geq 0$ contradicts Condition ii).

For the induction step (on n_F), we suppose that from i) – iii) it follows that the given C-ficograph is 2D embeddable if it has only k faces, for some $k \geq 1$. Now let $L = (V,H,I,c)$ satisfy i) – iii) and have k+1 faces.

By Lemma 2, there is an edge $e = \{h,h'\}$ in L such that $\mathrm{ang}(h) \in F_0$, but $\mathrm{ang}(h')$ belongs to a face $F_1 \neq F_0$. By Lemma 3, e is no bridge of L. Then the C-ficograph
$$L' = (V, H \setminus e, I \setminus \{(h,h'),(h',h)\}, c_{/H\setminus e})$$
has only k faces, namely the faces of L which are different from F_0 and F_1, and a face $\overline{F}$ corresponding to the fusion of F_0 and F_1. More precisely, if $F_0 = (h_0^0, v_0^0, h_0^{\cdot 0}, \ldots, h_1^0, v_1^0, h_1^{\cdot 0})$ and
$F_1 = (h_0^1, v_0^1, h_0^{\cdot 1}, \ldots, h_m^1, v_m^1, h_m^{\cdot 1})$, where $h_0^0 = h_m^{\cdot 1} = h$ and $h_0^1 = h_1^{\cdot 0} = h'$
(hence $v_0^0 = v_m^1$ and $v_1^0 = v_0^1$), then
$\overline{F} = (h_1^0, v_0^1 = v_1^0, h_0^{\cdot 1}, h_1^1, v_1^1, h_1^{\cdot 1}, \ldots, h_m^1, v_m^1 = v_0^0, h_0^{\cdot 0}, h_1^0, v_1^0, h_1^{\cdot 0}, \ldots$
$\ldots, h_{1-1}^0, v_{1-1}^0, h_{1-1}^{\cdot 0})$, cf. Fig. 13.

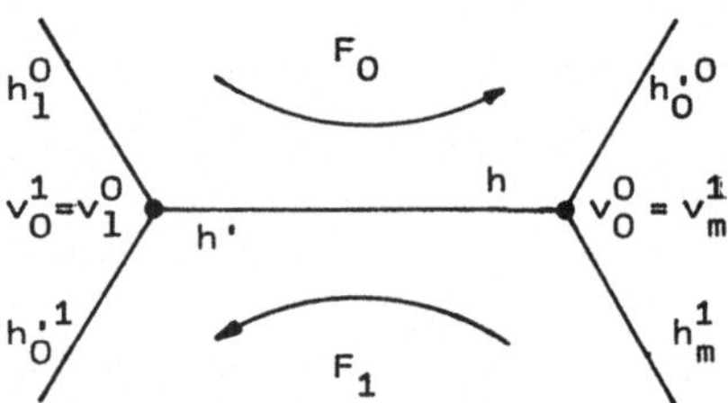

Fig. 13

It follows
$$\overline{\mathrm{rin}}(\overline{F}) = \overline{\mathrm{rin}}(F_0) + \overline{\mathrm{rin}}(F_1) - \mathrm{rin}(h_0^0, v_0^0, h_0^{\cdot 0})$$
$$- \mathrm{rin}(h_1^0, v_1^0, h_1^{\cdot 0}) - \mathrm{rin}(h_0^1, v_0^1, h_0^{\cdot 1}) - \mathrm{rin}(h_m^1, v_m^1, h_m^{\cdot 1})$$
$$+ \mathrm{rin}(h_1^0, v_0^1 = v_1^0, h_0^{\cdot 1}) + \mathrm{rin}(h_m^1, v_0^0 = v_m^1, h_0^{\cdot 0})$$
$$= 0 - 4 = -4,$$

as one proves by discussing all possible cases with respect to the rotation indices of the six angles occurring in the equation.

Therefore, L' satisfies the conditions i) – iii), and it is a 2D ficograph by the hypothesis of induction.

Now we define n_0 to be the smallest positive integer such that there are a 2D embedding P' of L' and a simple RPP

$w = (p_0, p_1, \ldots, p_{n_0})$,i.e. w consists of n_0+1 points,satisfying

$$(*) \begin{cases} p_0 = P'(v_0^0) \; , \; p_{n_0} = P'(v_0^1) \; , \\ \mathrm{dir}(\overrightarrow{p_0 p_1}) = \mathrm{dir}(\overrightarrow{p_{n-1} p_n}) = \overline{c(h)} = \overline{c(h')} \; , \text{ and} \\ (\bigcup_{i=0}^{n_0-1} \overline{p_i p_{i+1}} \smallsetminus \{p_0, p_{n_0}\}) \subseteq R_{\overline{F}} \; , \end{cases}$$

where $R_{\overline{F}}$ is the region corresponding to the face $\overline{F}$ with respect
to the embedding P' of L'.

<u>Claim.</u> $n_0 = 1$.

Obviously,this completes the proof of Proposition 2,since the
embedding P' and the simple RPP $w = (p_0, p_1)$ corresponding to
$n_0 = 1$ directly determine a 2D embedding P of the C-ficograph L.

It remains to prove the claim.

First we consider an arbitrary 2D embedding P' of L'. Since the
angles $(h_m^1, v_0^0 = v_m^1, h_0^{\cdot 0})$ and $(h_1^0, v_0^1 = v_1^0, h_0^{\cdot 1})$ belong to the same face $\overline{F}$
of L',there is a Jordan arc J connecting $P'(v_0^0)$ and $P'(v_0^1)$ and
leaving the region corresponding to $\overline{F}$ only in these points. Without loss of generality,we can assume that

$$J = \bigcup_{i=0}^{n-1} \overline{p_i p_{i+1}}$$

for some simple RPP $w = (p_0, p_1, \ldots, p_n)$ satisfying $(*)$ (for some
$n_0 = n$). Therefore,there is an integer n_0 with the above property.

Assume that $n_0 > 1$. Let P' and $w = (p_0, p_1, \ldots, p_{n_0})$ be the corresponding 2D embedding and the simple RPP,respectively. Since w is
simple,we have $r_i = \mathrm{rin}(p_i, p_{i+1}, p_{i+2}) \neq -2$ for $0 \leq i \leq n_0-2$. From the
minimality of n_0 it follows that $r_i \neq 0$.

If $\{r_i, r_{i+1}\} = \{1, -1\}$ for some $i \in \{0, \ldots, n_0-3\}$,we also obtain
a contradiction to the minimality of n_0. Indeed,if (without loss
of generality) $r_i = 1$, $r_{i+1} = -1$, and $\mathrm{dir}(\overrightarrow{p_i p_{i+1}}) = $ east,we have
a situation as in the proof of Lemma 5,see Fig. 10. Like in that
case,by the transformation $\mathcal{T}$ we obtain a 2D embedding $\hat{P}$ of L' and
a simple RPP $\hat{w} = (\hat{p}_0, \ldots, \hat{p}_{n_0-1})$ satisfying $(*)$.

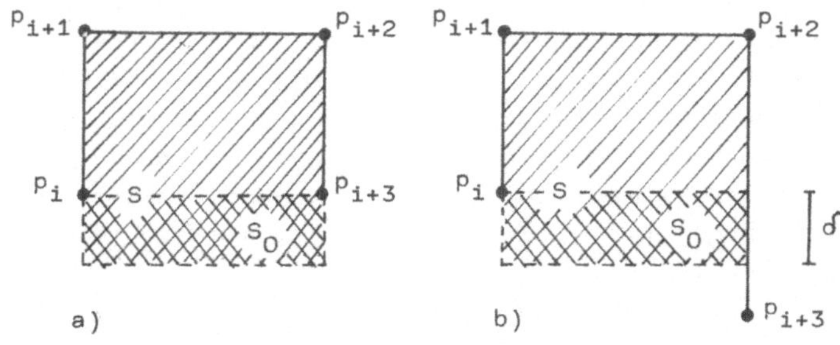

Fig. 14   a)   b)

Finally we consider the case that $r_i = r_{i+1} \in \{1,-1\}$ for some $i \in \{0, \ldots ,n_0-3\}$. Without loss of generality, let $r_i = r_{i+1} = -1$ and $\text{dir}(\overrightarrow{p_i p_{i+1}}) = \text{north}$. Fig. 14 a and b sketch the situations for $y_i = y_{i+3}$ and $y_i > y_{i+3}$, respectively, and rectangles S and S_0 such that there is a transformation $\mathcal{T}$ of S onto S_0 which leads to a contradiction to the minimality of n_0. The case $y_i < y_{i+3}$ is analogous. All notations are like those in the proof of Lemma 5.

This completes the proof of the claim and of Proposition 2. //

<u>Remark.</u> Conditions ii) and iii) in Proposition 2 can be replaced
   by
   ii') there is a face F_0 of L with $\overline{\text{rin}}(F_0) = -4$;
   iii') $\overline{\text{rin}}(F) \in \{4,-4\}$ for all faces F of L.

This easily follows from our proof. //

The C-ficograph in Fig. 15 shows that Condition ii') cannot be removed. Indeed, it has one face F_0 only, and $\overline{\text{rin}}(F_0) = 4$. Moreover, Condition i) is satisfied. In /L.Kr/ it is shown that ii') can be replaced by the condition that $\text{epc}(L) = 2$.

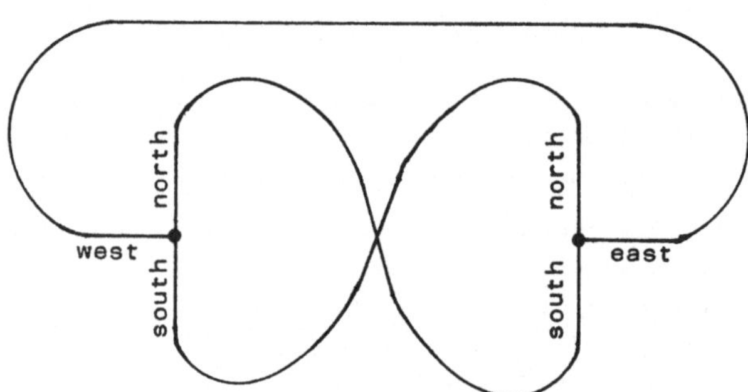

Fig. 15

<u>Hints & Sources.</u> The rotation index was introduced into labyrinth theory by L. Budach, see /L.Bu75,78a,b/. Using Budach's methods, K. Kriegel /L.Kr/ gave the elementary proof of Lemma 5, and F. Hoffmann and K. Kriegel /L.HoKr/,/L.Kr/ found Proposition 2. The notations used by these authors, however, differ from those introduced here. The importance of combinatorial characterizations and of 2D ficographs, generally, for layout problems was stressed by G. Vijayan and A. Wigderson /S.ViWi/. Also in computational geometry the rotation index (called sinuosity) of a rectilinear curve has been considered and applied to certain maze problems, see /S.SaSt/.

36

1.4. The hierarchy of types of labyrinths

The basic <u>types</u> <u>of</u> <u>labyrinths</u> introduced so far are summarized
in Fig. 16. Here the "gr" may stand for ("connected") "graphoid"
or "graph". If there is a directed path from one type to another,
then the latter one is or,in a natural way,can be regarded as a
subtype of the first.

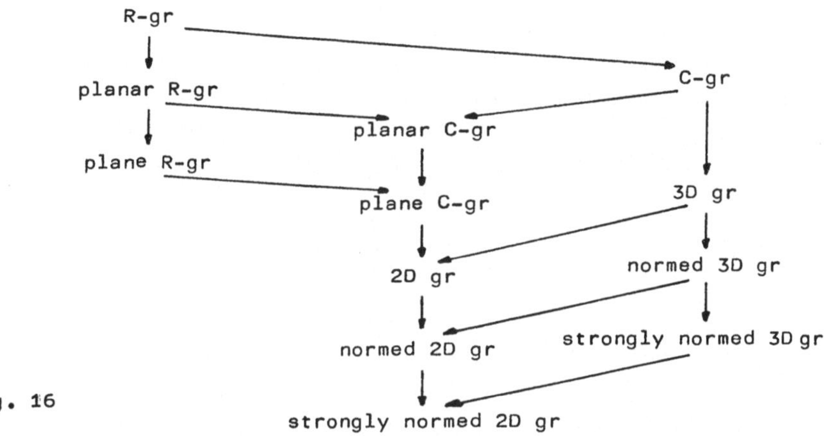

Fig. 16

This hierarchy of basic types of objects characterizes one
main component of labyrinth theory. The other one is given by the
("hierarchy of") types of automata which will be defined in the next
section. In this sense,labyrinth theory can be compared with
formal language theory (see /S.Sal/) in which it is an important
problem to characterize the complexity of types of languages by
determining the types of automata which are necessary and sufficient,respectively,to generate or recognize these types of languages.

The main types of languages are given by the well-known
Chomsky hierarchy consisting of four levels. Analogously,one can
select four main types of labyrinths. Fig. 17 shows this hierarchy.

We shortly recall some characteristic means which can be used
on the levels of the hierarchy from Fig. 16.

In R-graphoids one only has a cyclic permutation (rotation) of
the set of half-edges incident to a given vertex,in C-graphoids

this set is linearly ordered. In the planar case,we know that the
underlying graphoid is planar but the compass or rotation system
does not necessarily correspond to a plane embedding. This correspondence holds for plane R- or C-graphoids. Therefore,in walking
through such labyrinths,one can use Jordan's curve theorem and
its consequences.

In 3D graphoids the compass system corresponds to a 3D embedding,
analogously for 2D graphoids. In the latter type one can use the
rotation index of a path,for instance,in order to distinguish between interior and exterior faces. If a 3D graphoid is normed,then
one can compute the differences between the coordinates of
(embeddings of) vertices. In strongly normed 3D graphoids one can
even see whether or not there is a vertex with the distance 1 from
some current position in a certain direction.

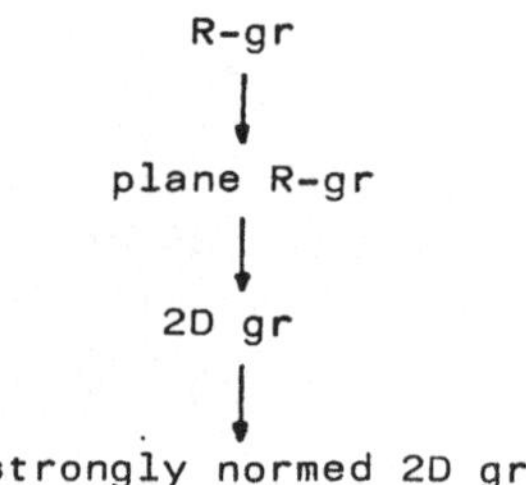

Fig. 17

Remark that the connected strongly normed 3D and 2D graphs
correspond to the 3- and 2-dimensional mazes,respectively.
Originally,an n-dimensional _maze_ is a connected set M of "cells"
of the discrete space Π^n. Two cells $z = (i_1,\ldots,i_n)$ and
$z' = (i_1',\ldots,i_n')$ are said to be neighbouring if

$$\sum_{j=1}^{n} (i_j-i_j')^2 = 1 ,$$ i.e.,they coincide in all components except

in exactly one in which they differ by 1. The connectedness of
sets $M \subseteq \Pi^n$ is straightforwardly defined. The example of a finite
2-dimensional maze M and the related 2D ficograph L(M) shown in
Fig. 18 a) and b) illustrates the strict correspondence between
these concepts.

The _order of connectivity_ of a (finite) 2-dimensional maze M
is defined to be the number of 8-connected components of the complement $\Pi^2 \setminus M$,see /S.Ro70/. The 8-connectedness is based on the
8-neighbourhood,and the cells $z = (i_1,i_2)$ and $z' = (i_1',i_2')$ are
called 8-neighbouring if $\max(|i_1-i_1'|,|i_2-i_2'|) = 1$. Thus,the order
of connectivity gives the number of 8-components of the barrier

of the maze.

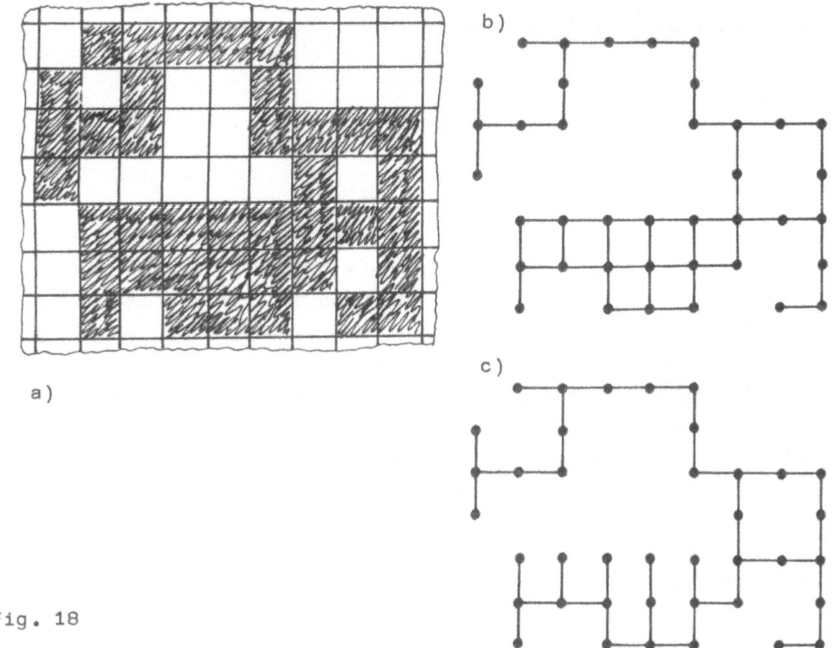

a)

b)

c)

Fig. 18

    For 2D ficographs L,it is natural to define the order of con-
nectivity as the number of faces of L. As Fig. 18 shows,then there
is no correspondence between the orders of connectivity of M and
L(M). Such a correspondence can be established if L(M) is reduced
in a suitable manner.

    If $L = (V,H,I,c)$ is a 2D ficograph,let
$$red(L) = (V,H_O , I_{/V \cup H_O} , c_{/H_O}) ,$$
where

$$H_O = H \smallsetminus \{ h: c(h) \in \{east,west\} \text{ and both ver(h) and}$$
$$verohal(h) \text{ have southern neighbours which}$$
$$\text{are connected by a horizontal edge} \} .$$

Fig. 18 c shows red(L(M)) for the sketched maze M. Without
giving the formal proof,we formulate

Lemma 7. For any finite 2-dimensional maze M, red(L(M)) is
    a 2D ficograph, and the number of faces of red(L(M))
    equals the order of connectivity of M. //

<u>Hints & Sources.</u> The idea to compare the known labyrinth results
and to put new questions by means of a hierarchy of the basic
labyrinth types was first published in /L.He85/.

<u>1.5. Automata in labyrinths</u>

In order to introduce the automata able to move in labyrinths,
we start with the simplest case of the finite automaton. At any
point of time such an automaton occupies a certain half-edge as
its position in the given labyrinth,and it can see the half-edges
incident to the corresponding vertex.

According to the basic types of labyrinths,we have to distinguish between finite R- and C-automata,respectively. In both cases
these are <u>Mealy</u> <u>automata</u> of the form

$$\alpha = (X , Y , A , \delta , \lambda , a_0),$$

where X is the input set,Y is the output set,A is the finite set
of states, $\delta : X \times A \longrightarrow A$ is the transition function, $\lambda : X \times A \longrightarrow Y$
is the output function,and $a_0 \in A$ is the initial state.

For R-graphoids we fix some degree bound $b \geqslant 3$. The automaton α
is called a <u>finite</u> <u>R-automaton</u> (of degree bound b) if

$$X = \left\{ 1,2,\ldots,b \right\} \quad \text{and} \quad Y = \left\{ \S,0,1,\ldots,b-1 \right\}.$$

This means the automaton can only see the number of half-edges
incident to the vertex ver(h) corresponding to its current position h. If the output information equals $\S$,it does not change its
position. If the output equals $y \in \left\{ 0,1,\ldots,b-1 \right\}$, then it moves
through the half-edge $r^y(h)$, this is the y th successor of h with
respect to the rotation in $H(\text{ver}(h))$.

More precisely,a <u>configuration</u> of α on an R-graphoid
$L = (V,H,I,r)$ is a pair $(h,a) \in H \times A$ giving the <u>current</u> <u>position</u> h
and the <u>current</u> <u>state</u> a of the automaton at some point of time.
Let

$$\delta (\text{deg}(\text{ver}(h)) , a) = a' \quad \text{and} \quad \lambda (\text{deg}(\text{ver}(h)) , a) = y.$$

If $y = \S$ and $h' = h$, or $y \neq \S$ and $h' = \text{hal} \circ r^y(h)$, then (h',a') is the
configuration of α in the next step; we also write

$$(h,a) \vdash^{\alpha} (h',a').$$

If $y \neq \S$ and $h_1 = r^y(h)$ is a free half-edge of L,we say that the
automaton <u>leaves</u> the graphoid (through the half-edge h_1,in the
state a') in the next step; we also write

$$(h,a) \downarrow^{\alpha} (h_1,a').$$

For C-graphoids we fix some direction set D. The automaton α is called a <u>finite</u> **C-automaton** (of direction set D) if
$$X = \{ (S,d): S \subseteq D \text{ and } d \in S \}, \quad Y = D \cup \{ \S \}, \text{ and}$$
$$\lambda((S,d),a) \in S \cup \{ \S \} \text{ for all } ((S,d),a) \in X \times A.$$
This means the automaton can see the directions in which there are half-edges incident to its current vertex, and the direction of its position, too. The output information is either $\S$, then it has the same position in the next step, or it gives the direction of the half-edge which has to be crossed.

More precisely, a <u>configuration</u> of α on a C-graphoid $L = (V,H,I,c)$ is a pair $(h,a) \in H \times A$. Let
$$\delta(c(H(ver(h))),a) = a' \quad \text{and} \quad \lambda(c(H(ver(h))),a) = d'.$$
If $d' = \S$ and $h' = h$, or $c(h_1) = d' \in D$, for $h_1 \in H(ver(h))$, and $h' = hal(h_1)$, then (h',a') is the next configuration; we write
$$(h,a) \vdash^{\alpha} (h',a').$$
If $c(h_1) = d' \in D$ for some free half-edge $h_1 \in H(ver(h))$, then we say that α <u>leaves</u> the C-graphoid L (through the half-edge h_1, in the state a'), and we write
$$(h,a) \downarrow^{\alpha} (h_1,a').$$
Let α be a (finite) R-automaton and L an R-graphoid or α be a **C**-automaton and L a C-graphoid of the corresponding degree bound and direction set, respectively. For any half-edge h of L, we define the <u>behaviour</u> of α in L, starting with position h. This is the infinite sequence
$$((h'_i,a'_i): i \in \mathbb{N}),$$
where $h'_0 = h$, $a'_0 = a_0$ and $(h'_i,a'_i) \vdash^{\alpha} (h'_{i+1},a'_{i+1})$ for $i \in \mathbb{N}$, if the automaton never leaves the graphoid. Otherwise, the behaviour is the finite sequence
$$((h'_0,a'_0),\ldots,(h'_l,a'_l)),$$
where $h'_0 = h$, $a'_0 = a_0$, $(h'_i,a'_i) \vdash^{\alpha} (h'_{i+1},a'_{i+1})$ for $0 \leq i < l-1$, and $(h'_{l-1},a'_{l-1}) \downarrow^{\alpha} (h'_l,a'_l)$.

We say that α <u>searches</u> the R- or C-graphoid L if, for any starting position h in L and any vertex v of L, the corresponding behaviour contains a position incident to v. We say that α <u>escapes</u> from L if, for any starting position, the corresponding behaviour is finite, i.e., the automaton always leaves the graphoid finally. Obviously, an automaton can only escape from open graphoids. Finally, we say that α <u>masters</u> the graphoid L if, for every starting position, the behaviour contains infinitely many positions. This is only possible if L is infinite. The problems of searching, escaping and mastering are fundamental in labyrinth theory.

Of course,the abilities of finite automata considered so far
are rather restricted. To obtain more powerful types of automata,
finite automata can be equipped with external auxiliary stores,
such as Turing worktapes,stacks,pushdown stores,counters etc.;
they may use special tokens,such as pebbles or pointers,to mark
elements of the given graphoid; they may form cooperating systems
of automata working simultaneously in the same graphoid; or they
may have multiple heads movable in the graphoid. Sometimes we
shall also consider combinations of these equipments,for instance,
automata using a pebble to mark vertices of the graphoid and
having an auxiliary counter additionally.

In the following we introduce these types of automata in more
detail but preferably in an informal manner. Only few representa-
tive examples of formal definitions are given to enable the reader
to make all definitions as formal and precise as he wants. If
necessary,he should consult /S.WaWe/ or /S.AhHoUl/ for this purpose. Throughout this book we prefer an informal,illustrative
treatment. So automata are mostly determined by programs describing their behaviour,not by formal definitions of their components.

A <u>Turing</u> <u>tape</u> <u>automaton</u> is a finite automaton equipped with a
Turing worktape and could be specified as a 7-tuple

$$\alpha = (X , Y , A , X_T , \delta , \lambda , a_0) ,$$

where X,Y,A and a_0 are defined as for a finite automaton,X_T is a
finite worktape alphabet containing a special blank symbol $\square$,
$\delta : X \times X_T \times A \longrightarrow X_T \times \{ 1,-1,\S \} \times A$, and $\lambda : X \times X_T \times A \longrightarrow Y$ with
$\lambda ((S,d) , x_T , a) \in S \cup \{ \S \}$ in the compass case. The second compo-
nent x_T of the arguments of the functions δ and λ is the inscription of the scanned cell of the worktape,and,by the first and the
second component of its values, δ gives the new inscription of
this scanned cell and the move instruction for the worktape head,
respectively.

A <u>configuration</u> of α on an R- resp. C-graphoid L is a quadruple (h,a,w_1,w_2) giving the position h of the automaton in the
labyrinth L , the internal state a , and the worktape inscriptions
$w_1,w_2 \in X_T^*$ to the left of the worktape head and to the right
inclusively the scanned cell,where infinite blank parts to the
left of w_1 and to the right of w_2 are omitted. The definitions of
the stepwise transition of configurations and of the behaviour of
the automaton on L are straightforward. Usually,we assume that the
automaton starts with a blank worktape.

Let $f: \mathbb{N}^+ \longrightarrow \mathbb{N}^+$. The Turing tape automaton α is said to be
$f(n)$ <u>space-bounded</u> on a certain type of labyrinths if,on any
labyrinth L of this type with exactly n vertices, α uses at most

O (f(n)) cells of its worktape.

The O-notation applied here means that the number of used cells can be bounded by $k \cdot f(n)$, where $k \in \mathbb{N}^+$ is a constant depending on α only. Generally, for functions $g, f : \mathbb{N}^+ \longrightarrow \mathbb{N}^+$ the notation

$$g(n) = O (f(n))$$

indicates that $g(n) \leq k \cdot f(n)$ for some constant $k \in \mathbb{N}^+$, or equivalently,

$$\lim_{n \to \infty} \inf \frac{f(n)}{g(n)} > 0 .$$

A stack automaton, pushdown automaton and counter automaton, respectively, can be regarded as a Turing tape automaton with a special way of working on its worktape. For a pushdown store, only a one-sided infinite part of the Turing worktape is used, and its first cell is marked by a special bottom symbol which cannot be changed by the automaton. The worktape head can only read the symbol on the top of the pushdown, can remove it from the pushdown (if it is not the bottom symbol), or can add a further non-blank symbol to the pushdown.

By a stack automaton, we mean an automaton with a Turing work-tape on which it operates like a pushdown automaton, i.e., it can recognize the bottom of the store, can pop the symbol on the top (if the store is non-empty), or it can push a further symbol into the store if the work head is on the top. But, additionally, the worktape head may scan all symbols within its store without changing them.

A counter automaton is a pushdown automaton whose worktape alphabet consists of the blank symbol $\square$, the bottom symbol and exactly one other symbol 1. Then the content of the store at any time represents a natural number (giving the number of 1-cells). This counter content can be increased and decreased, respectively, by one (the latter only if it is non-empty), and the automaton can recognize if its counter is empty.

Marker automata are finite automata equipped with a store of tokens able to indicate elements (half-edges or vertices) of the labyrinth. A token able to mark a vertex is called a pebble, and a token able to mark a half-edge is a pointer. We assume that the corresponding pebble or pointer automata can see the tokens indicating their positions or the vertices incident to their positions. They can pick up these tokens or deposit tokens on these places. We usually assume that any element of a graphoid can carry at most one token. Mostly, all pebbles and pointers, respectively, are indistinguishable for the automaton. But sometimes we shall also consider,for instance,k-pebble automata having k pebbles which are distinguishable each from the other or coloured,as one says then.

For instance,a <u>pointer automaton</u> can be specified as a Mealy automaton

$$\alpha = (\ \bar{X}\ ,\bar{Y}\ ,A\ ,\delta\ ,\lambda\ ,a_0\ ))\ ,$$

where

$$\bar{X} = X \times \{0,1\}\ ,\quad \bar{Y} = Y \times \{0,1\}\ ,$$

and X and Y are defined as for finite automata. Here the input (x,i) indicates the information $x \in X$ as in the case of a finite automaton but,additionally,that the position carries $i \in \{0,1\}$ pointers. Analogously,the output (y,j) means the move instruction $y \in Y$ as for a finite automaton but,moreover,that the automaton has to deposit $j \in \{0,1\}$ pointers on its last position. If $j = i$, we simply say that the marking at the position is not changed; if $i = 1$ and $j = 0$,we say that the automaton picks up the pointer from the position; and if $i = 0$ and $j = 1$,then it deposits a pointer on the position.

A <u>configuration</u> of the pointer automaton α on some R- or C-graphoid $L = (V,H,I,o)$ is a triple (h,a,p) , where $(h,a) \in H \times A$ gives the position and state of the automaton,and $p: H \longrightarrow \{0,1\}$ is a marker function yielding,for any half-edge $h \in H$,the number p(h) of pointers carried by h. The precise definitions of the transition function for configurations and of the behaviour of the automaton is straightforward. Of course,the automaton starts with the empty graphoid,i.e. without tokens in the graphoid.

Analogously one defines the concept of pebble automaton. Pointer and pebble automata using only $\leqslant k$ pointers and pebbles, respectively,in the course of their working are called <u>k-pointer automata</u> and <u>k-pebble automata</u>. Obviously,any k-pebble automaton can be simulated by a suitable k-pointer automaton.

Systems of k cooperating automata can be regarded as a general-ization of k-pointer automata. Here we have k finite automata which simultaneously work on the same labyrinth. Any two of these can see each other (i.e.,each can see the internal state of the other) if they have the same position. It is assumed that the automata can be distinguished each from the other and that each has its own program.

Formally,a <u>system of</u> k <u>cooperating automata</u> is a k-tuple

$$\gamma = (\ \alpha_1\ ,\ \alpha_2\ ,\ \ldots\ ,\alpha_k\ )$$

of Mealy automata

$$\alpha_i = (\ \bar{X}_i\ ,Y\ ,A_i\ ,\delta_i\ ,\lambda_i\ ,a_{0i}\ )\ ,\ 1 \leqslant i \leqslant k\ ,$$

with $\bar{X}_i = X \times (A_1 \cup \{\S\}) \times \ldots \times (A_{i-1} \cup \{\S\}) \times (A_{i+1} \cup \{\S\}) \times \ldots \times (A_k \cup \{\S\})$.

Here $A_1, \ldots, A_k$ are finite state sets, $\S \notin A_i$, $a_{0i} \in A_i$; X and Y
are defined as for finite automata.

The input $\overline{x}_i = (x, \overline{a}_1, \ldots, \overline{a}_{i-1}, \overline{a}_{i+1}, \ldots, \overline{a}_k)$ of the i th
automaton indicates the information x as in the case of finite
automata but,moreover,the states of all automata having the same
position as the i th one ($\overline{a}_j = \S$ means that the j th automaton has
another position).

A <u>configuration</u> of the system δ on some labyrinth L is a
2k-tuple

$$(h_1, a_1, h_2, a_2, \ldots, h_k, a_k)$$

giving the positions and states of the automata at some time. The
stepwise transition of configurations is sufficiently explained
by the above remarks. Usually we assume that all automata of the
system start on the same position. The given labyrinth is <u>searched</u>
if any vertex is visited by at least one automaton in the course
of the behaviour of the system starting on an arbitrary position.
We assume that the behaviour (as sequence) terminates if one of
the automata leaves the labyrinth through a free half-edge.

The requirement that the automata of a system can see each
other only if they have the same positions may seem to be rather
restrictive. But the concept does not become more powerful if one
allows that automata with neighbouring positions can see each
other. Indeed,systems of the second kind can be stepwise simulated
by such ones of the first kind.

A more powerful kind of cooperation is given by multihead automata which are quite common in complexity theory but have been
poorly investigated with respect to labyrinth problems so far.
A <u>k-head</u> <u>automaton</u> can be imagined as a system of k automata (the
heads) which interchange their input informations at any point of
time via a central control unit. Then the states of these automata
can be combined to one state of the whole system.

More precisely,a k-head automaton can be specified as a Mealy
automaton

$$\alpha = (\overline{X}, \overline{Y}, A, \delta, \lambda, a_0)$$

with $\overline{X} = X^k \times \mathcal{P}_k$ and $\overline{Y} = Y^k$, where X and Y are defined as for
finite automata of the corresponding type. Here $\mathcal{P}_k$ denotes the
set of all partitions of the set $\{1, \ldots, k\}$, i.e., $\mathcal{P}_k$ is the set
of all systems $P = \{P_1, \ldots, P_l\}$, where $1 \leq k$, $P_i \neq \emptyset$, $P_i \cap P_j = \emptyset$
for $1 \leq i, j \leq l$, $i \neq j$, and $P_1 \cup \ldots \cup P_l = \{1, \ldots, k\}$. The component
x_i of an input $\overline{x} = (x_1, \ldots, x_k, P)$ gives the input information of
the i th head like in the case of finite automata,whereas the

elements of the partition P are the "blocks" of heads having the
same position in the labyrinth.

We assume that the automaton starts in the initial state a_0
and with all heads on an arbitrary position, i.e. $P = \{\{1,...,k\}\}$
for the corresponding input $\bar{x}$. If any vertex is visited by some
head in the course of the working of the automaton, we say that the
given labyrinth is searched.

A jumping k-head automaton is still more comfortable. It is a
k-head automaton "each of whose heads can jump in a single step
to the current position of any other head", see /S.LeSe/. We omit
the formal details here.

We hope that the basic types of automata are sufficiently introduced by the remarks above. Moreover, the meaning of searching
labyrinths is clear now. Analogously, the concepts of mastering
labyrinths and escaping from labyrinths can be transferred to all
types of automata. Remark that for cooperating systems or multihead automata the definition of escaping from labyrinths involves
certain problems. It is possible to require that at least one
automaton or head leaves the given labyrinth or that all automata
or heads have to do this. Since we shall not deal with these problems in more detail, we omit such discussions.

We shall say that an automaton (or a system of cooperating
automata) halts at some time of its working if it does not change
its configuration in the next and all further steps. That means,
it does not change its position or the positions of all its automata or heads, it does not change the positions of its markers
(if there are any), and it does not change its internal state (or
states, in the case of a cooperating system). Equivalently, one
could require that the automaton enters a special stop state in
the corresponding step. But this would involve some problems for
cooperating systems.

Analogously, for cooperating systems the concept of recognition
or decision of properties (that are subclasses) of types of labyrinths involves problems. For an automaton of any other type introduced above we say that it recognizes a certain class $\mathcal{L}_0$ of
labyrinths of some type $\mathcal{L}$ if it, starting on an arbitrary position
in some labyrinth from $\mathcal{L}$, always halts with a special accepting
state if the labyrinth belongs to $\mathcal{L}_0$, and does not enter the accepting state otherwise. The automaton decides $\mathcal{L}_0$ with respect
to the type $\mathcal{L}$ if it, starting on an arbitrary position in a labyrinth from $\mathcal{L}$, always halts in a special accepting state if the
labyrinth belongs to $\mathcal{L}_0$ but halts in a rejecting state otherwise.

Besides the types of automata introduced so far,sometimes we also consider combinations of these,for instance (1-pebble,1-counter) automata. The meaning is straightforward.

Fig. 19 shows the hierarchy of basic types of automata. The arrow in $\mathcal{A}_1 \longrightarrow \mathcal{A}_2$ here indicates that the automata of type $\mathcal{A}_1$ are obviously able to simulate those of type $\mathcal{A}_2$, i.e., $\mathcal{A}_1$ is at least as powerful as $\mathcal{A}_2$. Of course,this relation is transitive.

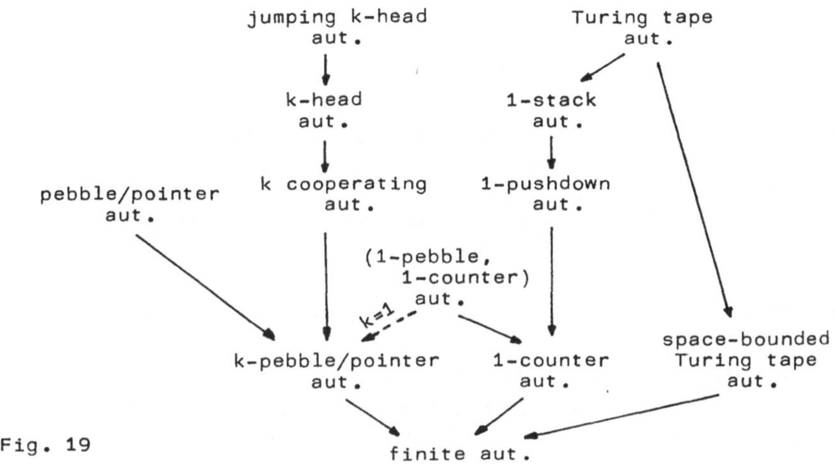

Fig. 19

So far we have only considered automata with at most one aux- iliary store (counter,pushdown,stack),since 2-counter automata are already computation universal,cf. Sections 1.6 and 2.8. Remark, however,that sometimes we shall also consider space-bounded counters,and the combination of two space-bounded counters,for instance,could be interesting.

<u>Hints & Sources.</u> Abstract automata like Turing machines which work on multidimensional tapes are as old as the basic concept of Turing machine. Finite C- and R-automata and the mastering problem for co-finite plane mazes and the searching problem for finite cubic R-graphs are essentially introduced by K. Döpp /L.Do/ and H. Müller /L.Mu71/,respectively. Independently,the searching problems of 2-dimensional patterns by several types of tape-bounded automata were considered by M. Blum and C. Hewitt /S.BlHe/, A. Rosenfeld and D. Milgram /S.MiRo/,and J. Mylopoulos /S.My/. Blum and Hewitt also investigated pebble automata on patterns. Pebble automata,cooperating systems and counter automata in

labyrinths were considered in publications by M. Blum,W. Sakoda
and D. Kozen /L.BlSa/,/L.BlKo/,but these authors refer to a seminar talk by M. Rabin (1967,unpublished). W. Coy /L.Co77,78/ published results on R-automata with auxiliary tapes. The slight
generalization of pebble automata to pointer automata was done in
/L.He84/. Finally,multihead automata are well-known from complexity theory,see /S.LeSe/; in connection with labyrinths similar
devices,namely jumping pebble controlling automata,were considered
by S. Cook and C. Rackoff /L.CoRa/.

1.6. Normal labyrinth problems and reducibility

A labyrinth problem is said to be <u>normal</u> if it can be characterized by a triple
$$\mathcal{P} = (\mathcal{L}, \mathcal{A}, \mathcal{T}),$$
where $\mathcal{L}$ is a class of R- or C-graphoids (a type of labyrinths),
$\mathcal{A}$ is a class of automata able to work in the labyrinths from $\mathcal{L}$
(type of automata),and $\mathcal{T}$ is a certain task to be done by some
automaton from $\mathcal{A}$ for all labyrinths of $\mathcal{L}$. Such a task may be
to search or to master the given labyrinths,to escape from them,
or to recognize or decide certain properties of them. Most problems of today's labyrinth theory are either normal or closely connected with normal problems.

A <u>solution</u> of the normal labyrinth problem $\mathcal{P}$ is an automaton
$\alpha \in \mathcal{A}$ which accomplishes the task $\mathcal{T}$ for all labyrinths $L \in \mathcal{L}$,
starting on an arbitrary position in L. Thus,a solution marks an
upper bound of the complexity of the task $\mathcal{T}$ on the labyrinth
type $\mathcal{L}$.

It may be that the problem is unsolvable,i.e.,the automata from
$\mathcal{A}$ are not powerful enough to accomplish the task $\mathcal{T}$ on the labyrinth type $\mathcal{L}$. In this way a lower bound of complexity is marked.
Then an effective proof of unsolvability,a so-called trap construction,is desirable. This is an effective procedure which,for any
automaton $\alpha \in \mathcal{A}$, yields a labyrinth $L \in \mathcal{L}$ and a starting position
h in L such that α does not accomplish the task $\mathcal{T}$ in L if it
starts on position h.

Let $\mathcal{P}_1 = (\mathcal{L}_1, \mathcal{A}, \mathcal{T})$ and $\mathcal{P}_2 = (\mathcal{L}_2, \mathcal{A}, \mathcal{T})$ be normal labyrinth problems such that $\mathcal{L}_1 \subseteq \mathcal{L}_2$. Then any solution of $\mathcal{P}_2$ is a
solution of $\mathcal{P}_1$, too. If $\mathcal{P}_1$ is unsolvable or there is a trap

construction for $\mathcal{P}_1$, then this is also the case for $\mathcal{P}_2$. In this
sense we say that solutions of normal problems are downwards her-
editary, but the unsolvability or trap constructions are upwards
hereditary in the hierarchy of types of labyrinths.

In order to explain the meaning of effectiveness used above in
more detail, we have to suppose that the automata of type $\mathcal{A}$ and
the labyrinths from $\mathcal{L}$ can be coded in some uniform and effective
way by words over a fixed alphabet. This is obvious for the types
of automata introduced in the previous section. For the types of
labyrinths introduced in Sections 1.2 and 1.3 there is no standard
code. Here we must presuppose that the elements of $\mathcal{L}$ can be
finitely characterized. This is possibly not the case for infinite
graphoids. But if we restrict ourselves to finite labyrinths or,
for instance, to co-finite 2-dimensional mazes, then there are
effective standard codes for the labyrinths L or even for the
pairs (L,h), where h is a position in L.

A normal labyrinth problem $\mathcal{P} = (\mathcal{L}, \mathcal{A}, \mathcal{T})$ is said to be
<u>codable</u> if the sets $\mathcal{A}$ and
$$\widehat{\mathcal{L}} = \{ (L,h) : L \in \mathcal{L} \text{ and } h \text{ is a position in L} \}$$
can be coded in some standard way by intuitively effective
one-to-one mappings
$$\varkappa_1 : \widehat{\mathcal{L}} \longrightarrow X_1^* \quad \text{and}$$
$$\varkappa_2 : \mathcal{A} \longrightarrow X_2^* ,$$
where X_1 and X_2 are suitable alphabets. (More precisely, $\varkappa_1$ and
$\varkappa_2$ are one-to-one only up to isomorphisms in $\widehat{\mathcal{L}}$ and $\mathcal{A}$, respectively.)

Then a <u>trap</u> <u>construction</u> for $\mathcal{P}$ is a recursive function f
satisfying the following two conditions.
(1) f($\varkappa_2(\alpha)$) is defined for all $\alpha \in \mathcal{A}$ and belongs to
     $\varkappa_1(\widehat{\mathcal{L}})$ then.
(2) If $\varkappa_1^{-1} \circ f \circ \varkappa_2(\alpha) = (L,h)$, then α does not accomplish
     the task $\mathcal{T}$ if it starts on position h in L.
  A codable problem $\mathcal{P}$ is called <u>effective</u> if
  - the domains $\varkappa_1(\widehat{\mathcal{L}})$ and $\varkappa_2(\mathcal{A})$ are recursively enumerable
    sets , and
  - the predicate
    $$\pi_{\mathcal{T}} (\varkappa_1(L,h), \varkappa_2(\alpha)) \equiv \text{" } \alpha \text{ accomplishes the task } \mathcal{T}$$
    $$\text{in L if it starts on h "}$$
    is recursively decidable with respect to the set $\varkappa_1(\widehat{\mathcal{L}}) \times \varkappa_2(\mathcal{A})$.

<u>Lemma 8.</u> If an effective normal labyrinth problem is not solvable,
     then there is a trap construction for it.

Let $\mathcal{P} = (\mathcal{L}, \mathcal{A}, \mathcal{T})$ be an unsolvable effective normal problem
with the code mappings $\varkappa_1$ and $\varkappa_2$. The algorithm defining the
recursive function f of the trap construction can work as follows.
Given a code $\varkappa_2(\alpha_0)$, for some $\alpha_0 \in \mathcal{A}$, it computes the value of
the predicate $\pi_{\mathcal{T}}(\varkappa_1(L,h), \varkappa_2(\alpha_0))$ for all $(L,h) \in \hat{\mathcal{L}}$ according to the enumeration procedure of $\varkappa_1(\hat{\mathcal{L}})$ as long as the value
TRUE is obtained. If $\pi_{\mathcal{T}}(\varkappa_1(L,h), \varkappa_2(\alpha_0)) = $ FALSE at the first
time, the algorithm halts with the result $\varkappa_1(L,h)$. //

The following notion of reducibility is introduced as a tool
to express certain simple relationships between normal labyrinth
problems. For normal problems $\mathcal{P} = (\mathcal{L}, \mathcal{A}, \mathcal{T})$ and $\mathcal{P}' = (\mathcal{L}', \mathcal{A}', \mathcal{T}')$,
we say that $\mathcal{P}$ is <u>reducible</u> to $\mathcal{P}'$ (briefly $\mathcal{P} \preccurlyeq \mathcal{P}'$) if any
solution α' of $\mathcal{P}'$ can be used to construct a solution α of $\mathcal{P}$.
More precisely, $\mathcal{P} \preccurlyeq \mathcal{P}'$ means that there is a recursive function f
such that if α' solves $\mathcal{P}'$, then $f \circ \varkappa'(\alpha')$ is defined, belongs to
$\varkappa(\mathcal{A})$, and $\varkappa^{-1} \circ f \circ \varkappa'(\alpha')$ is a solution of $\mathcal{P}$. Here let $\varkappa$ and
$\varkappa'$ denote some standard code mappings of the automaton types $\mathcal{A}$
and $\mathcal{A}'$, respectively.

If $\mathcal{P} \preccurlyeq \mathcal{P}'$, then to solve $\mathcal{P}$, it would be sufficient to solve $\mathcal{P}'$.
Conversely, if $\mathcal{P}'$ is unsolvable, then $\mathcal{P}$, too. Remark that an arbit-
rary problem can be reduced to any unsolvable problem by our
definition.

If $\mathcal{P} \preccurlyeq \mathcal{P}'$ and $\mathcal{P}' \preccurlyeq \mathcal{P}$, then the problems $\mathcal{P}$ and $\mathcal{P}'$ are said
to be <u>equivalent</u>, and we write $\mathcal{P} \equiv \mathcal{P}'$.

As a first simple example we note that a (k-) pebble automaton
can be simulated by a (k-) pointer automaton. Here the pointers
are used only to mark the vertices indicated by the pebbles of the
first automaton. It follows that a normal problem $(\mathcal{L}, \mathcal{A}, \mathcal{T})$, where
$\mathcal{A}$ is a class of (k-) pebble automata, can be reduced to $(\mathcal{L}, \mathcal{A}', \mathcal{T})$,
where $\mathcal{A}'$ is a class of (k-) pointer automata corresponding to $\mathcal{A}$.

Analogously, if the automata of some type $\mathcal{A}$ can be simulated
by automata from type $\mathcal{A}'$, e.g. as in the hierarchy shown in
Fig. 19, then any problem $(\mathcal{L}, \mathcal{A}, \mathcal{T})$ can be reduced to $(\mathcal{L}, \mathcal{A}', \mathcal{T})$.

Of course, strictly speaking the validity of the above assertions
depends both on the kind of the task $\mathcal{T}$ and on the kind of simulation. But we don't want to go into details on this general level
of dicussion; the assertions hold with respect to the tasks and
problems dealt with in this book.

It is well-known from automata theory that a Turing tape automaton can be simulated by a 2-pushdown automaton and even by a
2-counter automaton, and conversely. It follows the equivalence of
the problems $(\mathcal{L}, \mathcal{A}, \mathcal{T})$ and $(\mathcal{L}, \mathcal{A}', \mathcal{T})$, where $\mathcal{A}$ denotes a type

of Turing tape automata but $\mathcal{A}'$ is the corresponding type of 2-counter automata.

On C-graphoids a 1-pointer automaton can be simulated by a 1-pebble automaton. Here the pebble is used to mark the vertex v with which the pointer position is incident,the direction of the pointer position with respect to v is stored in a component of the state of the pebble automaton. Therefore,a normal problem concerning 1-pointer automata on a class of C-graphoids is equivalent to the corresponding problem concerning 1-pebble automata.

So far we have changed only the types of automata in the problems. Now we also modify the labyrinth types.

Since C-graphoids are special R-graphoids,any normal problem involving a type of C-graphoids can be reduced to a problem for a corresponding class of R-graphoids and with a corresponding type of R-automata.

In Section 1.4 we have shown that finite 2-dimensional mazes M correspond to 2D ficographs L(M). To obtain a coincidence between the order of connectivity of the maze and the number of faces of the corresponding 2D ficograph,we have assigned the ficograph red (L(M)) to the maze M. Any automaton α working in labyrinths red (L(M)) can obviously be simulated by an automaton $\tilde{\alpha}$ of the same type and working in the labyrinths L(M). Moreover,there is a recursive function f such that $\varkappa(\tilde{\alpha}) = f \circ \varkappa(\alpha)$ for some standard code mapping $\varkappa$.

Let $\mathcal{M}$ be a type of 2-dimensional mazes with an order of connectivity $\leqslant k$. Then any problem (L($\mathcal{M}$) , $\mathcal{A}$, $\mathcal{J}$) is reducible to (red $\circ$ L($\mathcal{M}$) , $\mathcal{A}$, $\mathcal{J}$) , and red $\circ$ L($\mathcal{M}$) only contains 2D ficographs with at most k faces. In other words,if we would have transferred the definitions of the basic types of automata from C-graphoids to 2-dimensional mazes,as one can straightforwardly do,then problems involving mazes with an order of connectivity $\leqslant k$ would be reducible to problems involving 2D ficographs with at most k faces. The types of automata would directly correspond one to the other.

A further intuitively simple reducibility result concerns the degree bound of R-graphoids. For $b \geqslant 3$ let $\mathcal{L}_b$ denote the class of all R-ficographs of degree bound b,and $\mathcal{A}_b$ denote a corresponding type of automata. Since $\mathcal{L}_3 \subseteq \mathcal{L}_b$, we have ($\mathcal{L}_3$, $\mathcal{A}_3$, $\mathcal{J}$) $\leqslant$ ($\mathcal{L}_b$, $\mathcal{A}_b$, $\mathcal{J}$).

On the other hand,by a simple vertex substitution σ corresponding to the rule sketched in Fig. 20,from any L $\in$ $\mathcal{L}_b$,we obtain a labyrinth σ (L) $\in \mathcal{L}_3$. (We shall deal with vertex substitutions in Section 1.8; these remarks should be understandable without further knowledge.) Often the work of any automaton $\alpha \in \mathcal{A}_3$ in a labyrinth $\sigma(L)$ can be simulated by an effectively constructible automaton $\alpha^{(\sigma)} \in \mathcal{A}_b$ working in $L \in \mathcal{L}_b$. Then we have $(\mathcal{L}_b, \mathcal{A}_b, \mathcal{J}) \leqslant (\mathcal{L}_3, \mathcal{A}_3, \mathcal{J})$, hence $(\mathcal{L}_3, \mathcal{A}_3, \mathcal{J}) \equiv (\mathcal{L}_b, \mathcal{A}_b, \mathcal{J})$.

The latter simulation may fail,for instance,in the case of pebble automata. Indeed,if α would simultaneously deposit pebbles on all deg(v) vertices in $\sigma(L)$ which are obtained by substitution from one vertex v of L,an automaton from $\mathcal{A}_b$ would not be able to simulate this (by means of indistinguishable pebbles).

Remark that the above considerations are also valid if we restrict $\mathcal{L}_b$ and $\mathcal{L}_3$ to plane labyrinths.

Fig. 20

Finally,we consider reducibilities between problems with different tasks.

Let $\mathcal{L}$ be the class of all infinite connected C-graphs and $\mathcal{L}_0$ be the class of all C-ficographs. An automaton α masters all labyrinths from $\mathcal{L}$ iff it searches all elements from $\mathcal{L}_0$. Indeed, if α does not master some $L \in \mathcal{L}$,then,starting at some position h in L,it becomes cyclic on a finite sublabyrinth L_0 of L. If $\overline{L_0}$ is the sublabyrinth of L,which is generated by L_0 and all neighbours of vertices of L_0,then α does not search $\overline{L_0}$,starting from position h. Conversely,if α does not search a certain $L_0 \in \mathcal{L}_0$,we consider a labyrinth $\overline{L_0} \in \mathcal{L}$ which is obtained from L_0 by connecting an edge e not crossed by α with an infinite chain of vertices, as sketched in Fig. 21. Obviously,$\overline{L_0}$ is not mastered by α.

Fig. 21

Therefore,the problem of searching the elements from $\mathcal{L}_0$ by an automaton of a certain type is equivalent to the problem of mastering the elements from $\mathcal{L}$ by an automaton of the same type. The analogous result holds if we consider the corresponding types of

infinite connected R-graphs and of all R-ficographs,respectively.
Moreover,we could take the corresponding plane labyrinths.

Now let $\mathcal{L}$ be the class of all infinite connected plane R- or
2D graphs without vertex accumulation points. The latter means
that there is a corresponding plane or 2D embedding such that in
any bounded region of the plane only finitely many vertices are
embedded. Let $\mathcal{L}_0$ denote the class of all plane R- and 2D ficographs,respectively.

Analogously to the above proof,one can show that the problem
of mastering the labyrinths from $\mathcal{L}$ is equivalent to the problem
of searching the exterior faces in the labyrinths from $\mathcal{L}_0$ by an
automaton of the given type. To search the exterior face means
that the automaton,starting on an arbitrary position,has to cross
all edges $e = \{h_1,h_2\}$ such that the half-edges h_1 and h_2 belong
to angles of the exterior face.

<u>Hints & Sources.</u> Further reducibility results will be discussed
or used later on. The idea of normal labyrinth problems is from
/L.He85/. Concerning such concepts as recursive function,recur-
sively decidable etc. the reader is referred to /S.WaWe/.

The following four sections deal with concepts and results
which will mainly be used in Chapter III. The reader who is preferably interested in searching algorithms can switch over to
Chapter II.

1.7. Finite automata in corridors

By an <u>R-</u> or <u>C-fragment</u>,we mean a tuple
$$F = (L, h_1, \ldots, h_l), \quad l \in \mathbb{N}^+,$$
where L is an R- or C-ficographoid,and the sequence $(h_1,\ldots,h_l)$
consists of all free half-edges of L and contains any of these
exactly once. F is said to be plane,normed plane,etc. if L has
the corresponding property.

Given a fragment $F = (L,h_1,\ldots,h_l)$ and a finite automaton
$\alpha = (X,Y,A,\delta,\lambda,a_0)$ of the corresponding type (R- or C-) , the
following function
$$\varphi_{\alpha,F} : A \times \{1,\ldots,l\} \longrightarrow (A \times \{1,\ldots,l\}) \cup \{\S\}$$

describes the <u>black-box</u> <u>behaviour</u> of α with respect to F considered from outside.

$$
\varphi_{\alpha,F}(a,i) = \begin{cases} (a',j) & \text{if } \alpha \text{, entering L through the half-edge } h_i \text{ in the state a, after finitely many steps leaves F through the half-edge } h_j \text{ in the state a' ,} \\ \S & \text{if } \alpha \text{, entering L through } h_i \text{ in the state a, never leaves this graphoid ,} \end{cases}
$$

where $(a,i) \in A \times \{1,\ldots,l\}$.

Fragments F_1 and F_2 of the same type and with the same number of free half-edges are said to be <u>α-equivalent</u> (briefly $F_1 \equiv_\alpha F_2$) if $\varphi_{\alpha,F_1} = \varphi_{\alpha,F_2}$.

An <u>isomorphism</u> between two such fragments is an R- or C-isomorphism between the underlying graphoids such that the linear order of free half-edges, as it is given by the corresponding sequence, is preserved.

Fragments with exactly two free half-edges,

$$F = (L,h_1,h_2) ,$$

are called <u>corridors</u>. The first half-edge h_1 is called the <u>entrance</u>, the second one, h_2, is the <u>exit</u> of F.

For corridors F and F' having no common elements (and being of the same type), the <u>concatenation</u> $F \cdot F'$ is the corridor obtained by interlinking the exit of F with the entrance of F'. More precisely, if $F = (L,h_1,h_2)$ and $F' = (L',h_1',h_2')$, then

$$F \cdot F' = (\hat{L}, h_1, h_2') , \text{where}$$
$$V_{\hat{L}} = V_L \cup V_{L'} ,$$
$$H_{\hat{L}} = H_L \cup H_{L'} ,$$
$$I_{\hat{L}} = I_L \cup I_{L'} \cup \{(h_1',h_2),(h_2,h_1')\} ,$$

and the rotation or compass system of $\hat{L}$ is obtained by joining the systems of L and L'.

Note that here, as also sometimes in the following, the components of a labyrinth, say L, are denoted by V_L, H_L, I_L and r_L or c_L.

Obviously, the concatenation is associative. Remark that the concatenation of 2D corridors is not necessarily a 2D corridor, since the directions of h_2 and h_1' may be non-opposite or h_1' and h_2 could belong to interior faces.

Naturally, the <u>power</u> F^i of a corridor F is defined to be the concatenation $F_1 \cdot F_2 \cdot \ldots \cdot F_i$ of i isomorphic copies $F_1,\ldots,F_i$ of F which have pairwise no common elements ($i \in \mathbb{N}^+$). F^i is uniquely determined up to isomorphism.

Lemma 9. Let α be a **finite** R- or C-automaton and F be a corridor of the corresponding type. For any state a of α the following holds.

If $\varphi_{\alpha,F}(a,1) \in (A \times \{1\}) \cup \{\S\}$, then $\varphi_{\alpha,F \cdot F'}(a,1) = \varphi_{\alpha,F}(a,1)$ for any corridor F' of the same type as F ;
analogously, if $\varphi_{\alpha,F}(a,2) \in (A \times \{2\}) \cup \{\S\}$, then $\varphi_{\alpha,F' \cdot F}(a,2) = \varphi_{\alpha,F}(a,2)$.

If α has only m states and $\varphi_{\alpha,F^m}(a,1) \in A \times \{2\}$, then $\varphi_{\alpha,F^{m+i}}(a,1) \in A \times \{2\}$, and from $\varphi_{\alpha,F^m}(a,2) \in A \times \{1\}$ it follows that $\varphi_{\alpha,F^{m+i}}(a,2) \in A \times \{1\}$, for all $i \in \mathbb{N}$.

The first two assertions hold trivially. We prove the third assertion, the fourth holds analogously.

Let α have only m states and $\varphi_{\alpha,F^m}(a,1) \in A \times \{2\}$, i.e., starting in state a at the entrance, the automaton leaves the corridor F^m through the exit after a certain number of steps. For $1 \leq i < m$, let a_i denote the state of the automaton in which it first crosses the edge connecting the i th with the (i+1) st copy of F in F^m. Accordingly, let $a_0 = a$ and $a_m = a'$ if $\varphi_{\alpha,F^m}(a,1) = (a',2)$, cf. Fig. 22.

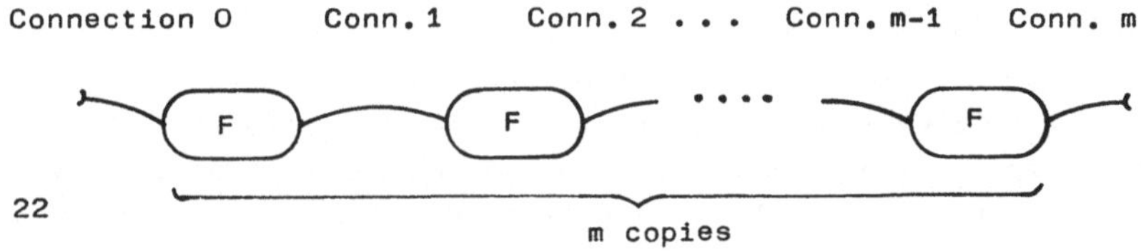

Connection 0   Conn. 1   Conn. 2 ...   Conn. m-1   Conn. m

Fig. 22

m copies

Since α has only m states, there are numbers i,j such that $0 \leq i < m$, $1 \leq j \leq m-i$ and $a_i = a_{i+j}$. Now let ($a_{i\,1} = a_i$, $a_{i\,2}$, ..., $a_{i\,l}$) be the sequence of states of α in which it crosses the i th connection before it reaches the (i+j) th one. Then l must be an odd number, and if $1 \leq 2k+1$ and $2k+2 < l$, we have

$$\varphi_{\alpha,F^j}(a_{i,2k+1},1) = (a_{i,2k+2},1) \quad \text{and}$$
$$\varphi_{\alpha,F^i}(a_{i,2k+2},2) = (a_{i,2k+3},2).$$

Moreover,

$$\varphi_{\alpha,F^j}(a_{i\,1},1) = \varphi_{\alpha,F^{i+j}}(a,1) = (a_{i+j},2) = (a_i,2).$$

It follows that the automaton, entering the corridor F^{i+2j} through the entrance with the state a, also crosses the (i+j) th connection in this corridor with the states $a_{i\,1} = a_i$, $a_{i\,2}$, ..., $a_{i\,l}$

and,after that,the $(i+2j)$ th connection in the state a_i. Therefore,
$$\varphi_{\alpha,F^{i+2j}}(a,1)=(a_i,2).$$
By induction we obtain
$$\varphi_{\alpha,F^{i+k\cdot j}}(a,1)=(a_i,2)\quad\text{for any }k\in\mathbb{N}^+.$$
For any $i'\in\mathbb{N}$,there is a $k\in\mathbb{N}^+$ such that $m+i'\leq i+k\cdot j$. Since $\varphi_{\alpha,F^{i+k\cdot j}}(a,1)\in A\times\{2\}$,we have $\varphi_{\alpha,F^{m+i'}}(a,1)\in A\times\{2\}$.//

<u>Proposition 3.</u> There is a function $\mu:\mathbb{N}^+\longrightarrow\mathbb{N}^+$ such that

$$\lim_{m\to\infty}\frac{\ln\circ\mu(m)}{\sqrt{m}\cdot\ln(m)}=2\text{ and the following holds.}$$

To any finite R- or C-automaton α with at most m states and to any corridor F of the corresponding type , there is a number $k\leq\mu(m)$ such that
$$F^k\equiv_\alpha F^{i\cdot k}\quad\text{for all }i\in\mathbb{N}^+.$$
Moreover,this equivalence always holds for
$$k=2\cdot\text{scm}\{2,3,\ldots,m\}.$$

By scm(S) ,we denote the smallest common multiple of a set S of natural numbers.

To prove the proposition,let be given a finite automaton $\alpha=(X,Y,A,\delta,\lambda,a_0)$ with at most m internal states,and an arbitrary corridor F of the corresponding type.

The behaviour of α in a corridor F^k,which it has entered through the entrance in a certain state a,can be described by crossing sequences as known from complexity theory.

For $1\leq i<k$ and $j\in\mathbb{N}^+$,let a_{ij} denote the state of α in which it crosses the i th connection between copies of F at the j th time after it has entered F^k through the entrance in the state a, cf. Fig. 22 for k = m. If Connection i is not crossed j times before the automaton leaves F^k, a_{ij} is undefined. Analogously,a_{0j} is the state in which α crosses the entrance half-edge and a_{kj} is the state in which it crosses the exit of F^k at the j th time. More precisely,

$$a_{01}=a,$$
$$a_{02}=\begin{cases}a' & \text{if }\varphi_{\alpha,F^k}(a,1)=(a',1),\\ \text{undefined} & \text{if }\varphi_{\alpha,F^k}(a,1)\notin A\times\{1\},\end{cases}$$
$$a_{0j}\text{ is undefined for }j\geq2,$$
$$a_{kj}\text{ is undefined for }j\geq1,\text{ and}$$
$$a_{k1}=\begin{cases}a' & \text{if }\varphi_{\alpha,F^k}(a,1)=(a',2),\\ \text{undefined} & \text{if }\varphi_{\alpha,F^k}(a,1)\notin A\times\{2\}.\end{cases}$$

For $0\leq i\leq k$,the sequence $(a_{i1},a_{i2},\ldots)$ consisting of all defined a_{ij} is called the i th <u>crossing sequence</u> (of α in F^k, starting at the entrance in the state a) .

Now let $k = 2m$.

<u>Case 1.</u> Starting in state a at the entrance, the automaton α never reaches Connection m in F^{2m}.

Then $\varphi_{\alpha, F^m}(a, 1) \in (A \times \{1\}) \cup \{\S\}$, hence

$$\varphi_{\alpha, F^{m+l}}(a, 1) = \varphi_{\alpha, F^m}(a, 1) \quad \text{for all } l \in \mathbb{N}.$$

<u>Case 2.</u> Starting in state a at the entrance, α reaches Connection m in F^{2m} before it leaves this corridor.

Then $\varphi_{\alpha, F^m}(a, 1) \in A \times \{2\}$, i.e., a_{m1} is defined (remember that $k = 2m$). Since α has at most m states, there are numbers i_0, p such that

$$0 \leq i_0 < m, \quad 1 \leq p \leq m - i_0 \quad \text{and} \quad a_{i_0 1} = a_{i_0 + p, 1}.$$

We prove that the i_0 th and the $(i_0 + p)$ th crossing sequence are equal.

<u>Claim 1.</u> For any $j \in \mathbb{N}^+$ it holds that $a_{i_0 j}$ is defined iff $a_{i_0 + p, j}$ is defined, and $a_{i_0 j} = a_{i_0 + p, j}$ in this case.

The claim holds for $j = 1$. Supposing its validity for j, we show it for $j + 1$.

This trivially holds if $a_{i_0 j}$ and $a_{i_0 + p, j}$ are undefined. Now let $a_{i_0 j}$ and $a_{i_0 + p, j}$ be defined.

Assume first that j is odd. Then the j th crossings of Connections i_0 and $i_0 + p$ are from left to right. If $a_{i_0, j+1}$ is defined, then $\varphi_{\alpha, F^{2m - i_0}}(a_{i_0 j}, 1) \in A \times \{1\}$, and we have

$$\varphi_{\alpha, F^{2m - i_0}}(a_{i_0 j}, 1) = (a_{i_0, j+1}, 1).$$

Since $2m - i_0 - p \geq m$, from Lemma 9 it follows that $\varphi_{\alpha, F^{2m - i_0 - p}}(a_{i_0 j}, 1) \in A \times \{1\}$ and $\varphi_{\alpha, F^{2m - i_0 - p}}(a_{i_0 j}, 1) = (a_{i_0, j+1}, 1)$. Hence $a_{i_0 + p, j+1}$ is defined and equal to $a_{i_0, j+1}$.

If $a_{i_0, j+1}$ is undefined, hence $\varphi_{\alpha, F^{2m - i_0}}(a_{i_0 j}, 1) \in (A \times \{2\}) \cup \{\S\}$, then it follows that $\varphi_{\alpha, F^{2m - i_0 - p}}(a_{i_0 j}, 1) \in (A \times \{2\}) \cup \{\S\}$. This means that $a_{i_0 + p, j+1}$ is undefined.

Now we assume that j is an even number. Therefore, the j th crossings of Connections i_0 and $i_0 + p$ are from right to left. Since a_{m1} is defined, by Lemma 9 we obtain that $\varphi_{\alpha, F^{2m}}(a, 1) \in A \times \{2\}$. Hence $a_{i_0, j+1}$ must be defined, and

$$\varphi_{\alpha, F^{i_0}}(a_{i_0 j}, 2) = (a_{i_0, j+1}, 2).$$

It follows that $\varphi_{\alpha, F^{i_0+p}}(a_{i_0 j}, 2) = (a_{i_0, j+1}, 2)$, and
$a_{i_0+p, j+1} = a_{i_0, j+1}$. The claim has been proved.

From the equality of the i_0 th to the (i_0+p) th crossing sequence, it follows that
$$\varphi_{\alpha, F^{2m}}(a, 1) = \varphi_{\alpha, F^{2m-p}}(a, 1).$$
The proof of Claim 1 can immediately be transferred to obtain the more general

<u>Claim 2.</u> Assume that $\varphi_{\alpha, F^m}(a, 1) \in A \times \{2\}$, and let the automaton α cross the Connections i and $i+p$ in the power F^k with the same state at the first time, for some k satisfying $k-i-p \geqslant m$. Then the i th and the $(i+p)$ th crossing sequence (of α in F^k, starting with the state a at the entrance) coincide.

Moreover, under the suppositions of Claim 2 it follows that also the $(i+1)$ th and the $(i+p+1)$ th crossing sequence coincide as long as $k-i-p-1 \geqslant m$, and it holds
$$\varphi_{\alpha, F^{k-1}}(a, 1) = \varphi_{\alpha, F^{k-p-1}}(a, 1)$$
in this case.

To any state $a \in A$ which satisfies $\varphi_{\alpha, F^{2m}}(a, 1) \in A \times \{2\}$, we assign the uniquely determined cyclic permutation
$$Z_a = (a_{i_0 1}, a_{i_0+1, 1}, \cdots, a_{i_0+p_a-1, 1})$$
of the first states of the crossing sequences of Connections i_0, $i_0+1, \ldots, i_0+p_a+1$ in F^{2m}, where $0 \leqslant i_0 < m$, $1 \leqslant p_a \leqslant m-i_0$, $a_{i_0 1} = a_{i_0+p_a, 1}$, and p_a is minimal with these properties.

Then
$$\varphi_{\alpha, F^k}(a, 1) = \varphi_{\alpha, F^{k-p_a}}(a, 1) \quad \text{for } k \geqslant 2m.$$
From $Z_{a_1} \neq Z_{a_2}$, for $a_1, a_2 \in A$ with $\varphi_{\alpha, F^{2m}}(a_i, 1) \in A \times \{2\}$
$(i = 1, 2)$, it follows that Z_{a_1} and Z_{a_2} have no common elements.

Therefore, if we define
$$P_\alpha = \{p_a : a \in A \text{ and } \varphi_{\alpha, F^{2m}}(a, 1) \in A \times \{2\}\},$$
then the sum of all elements of P_α is not graeter than m,
$$\sum_{(p \in P_\alpha)} p \leqslant m.$$
Let $k_1 = \text{scm}(P_\alpha)$. We obtain
$$\varphi_{\alpha, F^k}(a, 1) = \varphi_{\alpha, F^{k+k_1}}(a, 1)$$
or any $a \in A$ and $k \geqslant 2m$.

Analogously, investigating the behaviuor of α entering the corridors F^k for $k \geqslant 2m$ through the exit (i.e. from the right-hand ide), we obtain a set

$$P'_\alpha = \left\{ p'_a : a \in A \text{ and } \varphi_{\alpha, F^{2m}}(a,2) \in A \times \{1\} \right\}$$

such that

$$\sum_{(p \in P'_\alpha)} p \leqslant m$$

and, for $k_r = \mathrm{scm}(P'_\alpha)$, $a \in A$ and $k \geqslant 2m$,

$$\varphi_{\alpha, F^k}(a,2) = \varphi_{\alpha, F^{k+k_r}}(a,2).$$

In /S.La/, pp. 222-229, E. Landau estimated the function

$$\gamma(m) = \max \left\{ \mathrm{scm}(S) : S \subseteq \mathbb{N}^+ \text{ and } \sum_{(s \in S)} s \leqslant m \right\}$$

and proved that

$$\lim_{m \to \infty} \frac{\ln \circ \gamma(m)}{\sqrt{m \cdot \ln(m)}} = 1.$$

Now let

$$k^* = j^* \cdot k_1 \cdot k_r \,,$$

where j^* is the minimal number $j \in \mathbb{N}^+$ such that

$$j \cdot k_1 \cdot k_r \geqslant 2m.$$

Then

$$\varphi_{\alpha, F^{k^*}}(a,1) = \varphi_{\alpha, F^{i \cdot k^*}}(a,1)$$

for any $a \in A$, $i \in \mathbb{N}^+$ and $l = 1,2$; that means

$$F^{k^*} \equiv_\alpha F^{i \cdot k^*}.$$

Moreover, for

$$\mu(m) = 2m \cdot \gamma^2(m),$$

we have $k^* \leqslant \mu(m)$ and

$$\lim_{m \to \infty} \frac{\ln \circ \mu(m)}{\sqrt{m \cdot \ln(m)}} = 2.$$

The first part of Proposition 3 has been proved.

To show the second part, we remark that for

$$k^+ = 2 \cdot \mathrm{scm}\{2, \ldots, m\}$$

we have $k^+ \geqslant 2m$, $k^+ = j_1 \cdot k_1$ and $k^+ = j_r \cdot k_r$ with suitable numbers $j_1, j_r \in \mathbb{N}^+$. It follows

$$F^{k^+} \equiv_\alpha F^{i \cdot k^+} \qquad \text{for all } i \in \mathbb{N}^+. \quad //$$

<u>Hints & Sources.</u> This section is a straightforward generalization of results known for special (plane) C-corridors whose underlying graphoids have degree 2. In this sense, the second part of Proposition 3 and the basic ideas are due to L. Budach /L.Bu75,78a,b/. The first part and especially the reference to Landau's result were contributed by H. Antelmann /L.An/. The description of the black-box behaviour of a finite automaton α on a fragment by a function and the corresponding concept of α-equivalence were already used by M. Blum and C. Hewitt /S.BlHe/.

1.8. Vertex substitutions

The methods of vertex and edge substitution,respectively,which
we shall consider now are the basis for certain simulation techniques. Vertex substitutions replace the vertices together with
their incident half-edges in labyrinths of a certain type by so-
called hypervertices,these are suitable R- or C-graphoids. Edge
substitutions are based on insertions of suitable corridors into
the edges of labyrinths. The basic ideas of these methods are
rather simple,nevertheless,an exact treatment requires some effort.

By a _star_ , we mean a labyrinth (i.e. a connected R- or C-
graphoid) with only one vertex and some half-edges all incident to
this vertex but not incident among one another. Accordingly,the
star _of_ _vertex_ v in a labyrinth L is the following sublabyrinth
of L,

$$S_L(v) = (\{v\}, H_L(v), \{(v,h), (h,v) : h \in H_L(v)\}, o_{L,v}),$$

where $o_{L,v}$ stands for the restriction of the rotation or compass
system of L to $H_L(v)$. In other words, $S_L(v)$ is the sublabyrinth
generated by the elements from $\{v\} \cup H_L(v)$.

A _vertex_ _substitution_ on a labyrinth type $\mathcal{L}$ is a finite system
of rules,

$$(\ast) \qquad \sigma = \left\{(S_i, h_{i1}, \ldots, h_{i\,l_i}) \longrightarrow (L'_i, h'_{i1}, \ldots, h'_{i\,l_i}) : i = 1, \ldots, k\right\},$$

$k \in \mathbb{N}^+$, where the following properties hold .

(1) Both the left-hand sides $F_i = (S_i, h_{i\,1}, \ldots, h_{i\,l_i})$ and the

right-hand sides $F'_i = (L'_i, h'_{i\,1}, \ldots, h'_{i\,l_i})$ of the rules

are either R- or C-fragments ; the R- or C-graphoids S_i are

isomorphic to stars of vertices in labyrinths from $\mathcal{L}$.

(2) To any star $S_L(v)$ of a vertex v in a labyrinth $L \in \mathcal{L}$,there

is exactly one rule $F_{i(L,v)} \longrightarrow F'_{i(L,v)}$ such that the cor-

responding star $S_{i(L,v)}$ is isomorphic to $S_L(v)$.

(3) a) If $\mathcal{L}$ is a class of R-graphoids,then the R-fragments F'_i

must be _rotationally_ _symmetric_ ; that means,for any

rotation r' of the set $\{1,2,\ldots,l_i\}$, the R-fragments F'_i

and $(L'_i, h'_{i\,r'(1)}, \ldots, h'_{i\,r'(l_i)})$ are isomorphic .

b) If $\mathcal{L}$ is a class of C-graphoids,then the corresponding

free half-edges on both sides of any rule are marked by

the same directions ; i.e. $c_{L'_i}(h'_{i\,j}) = c_{S_i}(h_{i\,j})$ for

$1 \leqslant j \leqslant l_i$.

Concerning the property (2) **we remark** that if the degrees of
labyrinths from $\mathcal{L}$ are universally bounded by some $b \geqslant 3$ (as usual
for our purposes),then there are only finitely many mutually non-
isomorphic stars of vertices in labyrinths from $\mathcal{L}$.

Given a labyrinth $L \in \mathcal{L}$,we obtain the labyrinth $\sigma(L)$ by
replacing all stars $S_L(v)$ in L by (mutually disjoint) isomorphic
copies of the graphoids $L'_{i(L,v)}$ on the right-hand sides of the
corresponding rules of σ. More precisely,for any vertex $v \in V_L$,let
$$(L'_v , h'_{v\,1} , \ldots , h'_{v\,l_v}) \quad \text{with} \quad L'_v = (V_v,H_v,I_v,o_v)$$
be an isomorphic copy of the fragment
$$F'_{i(L,v)} = (L'_{i(L,v)} , h'_{i(L,v)} , \ldots , h'_{i(L,v)} , l_{i(L,v)})$$
such that L'_{v_1} and L'_{v_2} have no common elements if $v_1 \neq v_2$. Then
$$\sigma(L) = (\overline{V} , \overline{H} , \overline{I} , \overline{o})$$
with
$$\overline{V} = \bigcup_{v \in V_L} V_v \, ,$$

$$\overline{H} = \bigcup_{v \in V_L} H_v \, ,$$

$$\overline{o} = \bigcup_{v \in V_L} o_v \, ,$$

$$\overline{I} = \bigcup_{v \in V_L} I_v \, \cup \, \widehat{I} \, ,$$
where
$$\widehat{I} = \Big\{ (h'_{v_1\,j_1} , h'_{v_2\,j_2}) : v_1,v_2 \in V_L \text{ and } (h_{v_1\,j_1},h_{v_2\,j_2}) \in I_L ,$$
$$\text{where } (S_L(v_p),h_{v_p\,1} ,\ldots,h_{v_p\,l_{v_p}})$$
$$\text{are the chosen fragments}$$
$$\text{isomorphic to}$$
$$(S_{i(L,v_p)},h_{i(L,v_p)}, 1 , \ldots$$
$$\ldots , h_{i(L,v_p)} , l_{i(L,v_p)}),$$
$$\text{for } p = 1 , 2 \Big\} .$$

In this way, $\sigma(L)$ is uniquely determined up to isomorphism.
The graphoids L'_v are called the <u>hypervertices</u> of $\sigma(L)$, the
edges defined by I give the connections between hypervertices.

To explain the phrase "chosen fragments" in the definition of
I , we remark that in the case of R-graphoids L the fragments
$(S_L(v) , h_{v , r'(1)} , \ldots , h_{v , r'(l_v)})$ are isomorphic each to the
other for any rotation r' of $\{1,\ldots,l_v\}$. To define I , we have to
choose exactly one member of this isoclass. By Property (3a),this

choice does not influence the resulting labyrinth $\sigma(L)$ up to
isomorphism. In the case of C-fragments, this problem does not
arise, since any half-edge $h_{v\,i} \in H_L(v)$ is uniquely distinguishable
from the other elements of $H_L(v)$ by its direction.

Fig. 23 shows an example of a vertex substitution σ for R-
graphoids of a degree bound $b \geqslant 3$. The obtained labyrinths $\sigma(L)$
are cubic, and from plane R-graphoids L we obtain plane labyrinths
$\sigma(L)$ by this substitution.

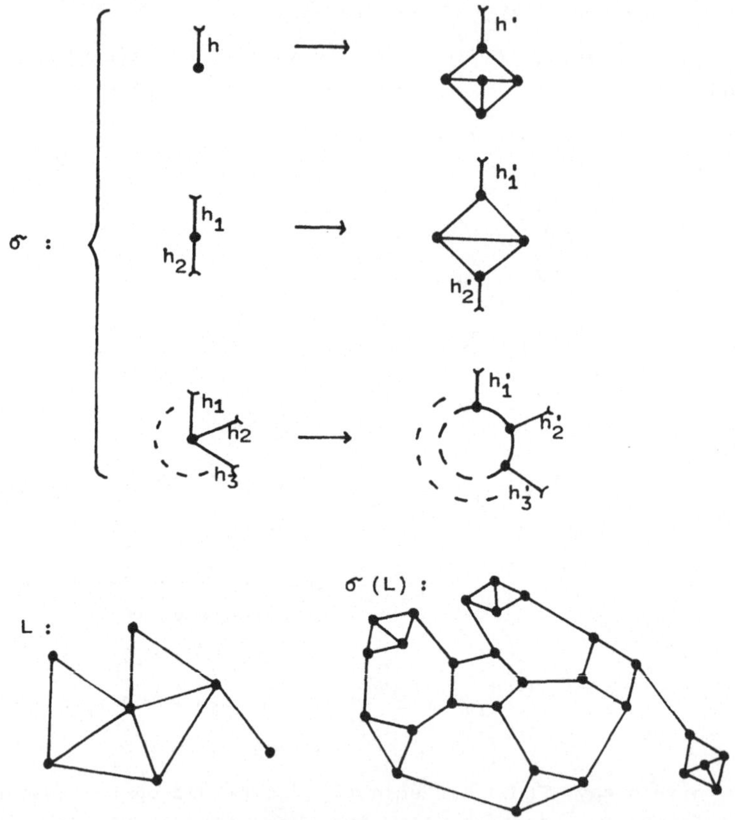

Fig. 23

Fig. 24 shows a vertex substitution for 2D graphoids. It replaces all vertices of degree four by hypervertices of degree
three. Vertices of other degrees remain unchanged.

Let be given a vertex substitution σ on some labyrinth type $\mathcal{L}$
and a finite automaton $\mathcal{O}$ working on all labyrinths $\sigma(L)$ for $L \in \mathcal{L}$.

There is a canonical way to construct a finite automaton $\mathfrak{A}(\sigma)$ which, working on a labyrinth $L \in \mathcal{L}$, simulates the behaviour of $\mathfrak{A}$ working on $\sigma(L)$. Moreover, the starting positions of the simulated behaviour can be arbitrarily fixed within the hypervertices corresponding to the starting vertices of $\mathfrak{A}(\sigma)$.

σ :

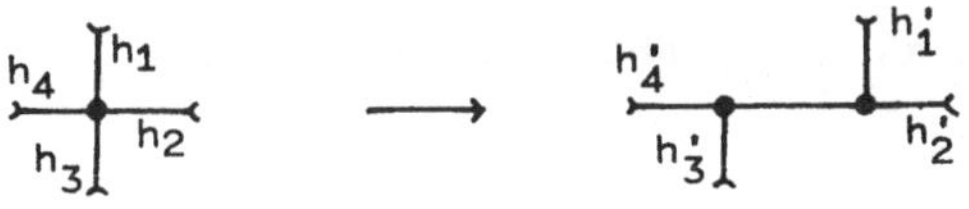

identical transformations for vertices of degrees ≤ 3

Fig. 24

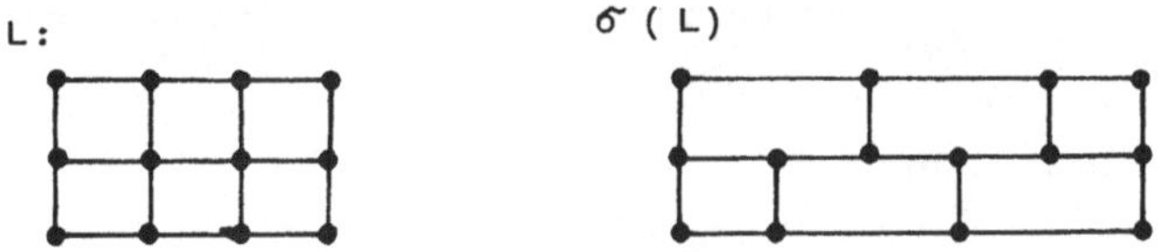

Let σ have the form (*) and $h'_{i\,0}$ be fixed half-edges in the graphoids L'_i for $1 \leq i \leq k$. If $\mathfrak{A} = (X, Y, A, \delta, \lambda, a_0)$, then we define

$$\mathfrak{A}(\sigma) = (\overline{X}, \overline{Y}, \overline{A}, \overline{\delta}, \overline{\lambda}, \overline{a}_0)$$

as follows.

The alphabets $\overline{X}$ and $\overline{Y}$ corrospond to the labyrinth type $\mathcal{L}$; especially if $\mathcal{L}$ consists of C-graphoids, then $\sigma(\mathcal{L})$ too, and $\overline{X} = X$, $\overline{Y} = Y$.

We define

$$\overline{A} = \{\overline{a}_0\} \cup (\{0\} \times A) \cup \left(\left(\bigcup_{i=1}^{k} H_{L'_i} \right) \times A \right),$$

where the three components of this union be pairwise disjoint.

To explain the definition of $\overline{\delta}$ and $\overline{\lambda}$, we remark that any input symbol $\overline{x} \in \overline{X}$ uniquely determines an index $i(\overline{x}) \in \{1, \ldots, k\}$ such that the star $S_{i(\overline{x})}$ is isomorphic to the star $S_L(ver(h))$ of any position h on which the input $\overline{x}$ occurs in a labyrinth $L \in \mathcal{L}$. Moreover, let $j(\overline{x}) \in \{1, \ldots, l_{i(\overline{x})}\}$ such that the half-edge $h_{i(\overline{x}),j(\overline{x})}$ in $S_{i(\overline{x})}$ corresponds to the position h at vertex ver(h). In the C-case, $j(\overline{x})$ is uniquely determined by the direction in the second component of $\overline{x}$; in the R-case, $j(\overline{x})$ can be arbitrarily chosen from $\{1, \ldots, l_{i(\overline{x})}\}$, and, by the property (3a), this choice does not influence the result up to isomorphism.

For any $\bar{x} \in \bar{X}$, let
$$\bar{\sigma}(\bar{x},\bar{a}_0) = (h'_{i(\bar{x})0}, a_0), \quad \bar{\lambda}(\bar{x},\bar{a}_0) = \S.$$

If $h'_1, h'_2 \in H_{L_{i(\bar{x})}}$ and $(h'_1, a_1) \overset{\alpha}{\longmapsto} (h'_2, a_2)$,

then let
$$\bar{\sigma}(\bar{x},(h'_1,a_1)) = (h'_2,a_2), \quad \bar{\lambda}(\bar{x},(h'_1,a_1)) = \S.$$

For $h'_1 \in H_{L_{i(\bar{x})}}$ and $(h'_1, a_1) \overset{\alpha}{\downarrow} (h'_{i(\bar{x}),j_1}, a_2)$,

we define
$$\bar{\sigma}(\bar{x},(h'_1,a_1)) = (0, a_2), \quad \text{and}$$
$$\bar{\lambda}(\bar{x},(h'_1,a_1)) = \bar{y}(h_{i(\bar{x}),j_1}),$$

where $\bar{y}(h_{i(\bar{x}),j_1})$ be the output symbol causing an automaton

that occupies the position $h_{i(\bar{x}),j(\bar{x})}$ to move through the half-

edge $h_{i(\bar{x}),j_1}$ corresponding in $F_{i(\bar{x})}$ to the half-edge $h'_{i(\bar{x}),j_1}$

in $F'_{i(\bar{x})}$.

Finally, let
$$\bar{\sigma}(\bar{x},(0,a_1)) = (h'_{i(\bar{x}),j(\bar{x})}, a_1),$$
$$\bar{\lambda}(\bar{x},(0,a_1)) = \S.$$

Disregarded of the considerable formal effort, one easily shows

<u>Lemma 10.</u> Let σ be a vertex substitution on a labyrinth type $\mathcal{L}$,
   and α be a finite automaton able to work on the labyrinths
   from $\mathcal{L}$.

   Then, on any labyrinth $L \in \mathcal{L}$, it holds

$$(h_1, \bar{a}_0) \overset{\alpha(\sigma)}{\longmapsto} (h_1, (h'_{i_1 0}, a_0)) \overset{\alpha(\sigma)}{\longmapsto}$$
$$(h_1, (h'_{i_1 1}, a_{11})) \overset{\alpha(\sigma)}{\longmapsto} \ldots \overset{\alpha(\sigma)}{\longmapsto} (h_1, (h'_{i_1 k_1}, a_{1 k_1})) \overset{\alpha(\sigma)}{\longmapsto}$$
$$(h_2, (0, a_2)) \overset{\alpha(\sigma)}{\longmapsto} (h_2, (h'_{i_2 1}, a_2)) \overset{\alpha(\sigma)}{\longmapsto}$$
$$(h_2, (h'_{i_2 2}, a_{22})) \overset{\alpha(\sigma)}{\longmapsto} \ldots \overset{\alpha(\sigma)}{\longmapsto} (h_2, (h'_{i_2 k_2}, a_{2 k_2})) \overset{\alpha(\sigma)}{\longmapsto}$$
$$(h_3, (0, a_3)) \overset{\alpha(\sigma)}{\longmapsto} (h_3, (h'_{i_3 1}, a_3)) \overset{\alpha(\sigma)}{\longmapsto}$$
$$(h_3, (h'_{i_3 2}, a_{32})) \overset{\alpha(\sigma)}{\longmapsto} \ldots \ldots$$

   iff, on the labyrinth $\sigma(L)$,

$$(+) \quad \begin{cases} (\varphi_{h_1}(h'_{i_1 0}), a_0) \overset{\alpha}{\longmapsto} \\[2mm] (\varphi_{h_1}(h'_{i_1 1}), a_{11}) \overset{\alpha}{\longmapsto} \ldots \overset{\alpha}{\longmapsto} (\varphi_{h_1}(h'_{i_1 k_1}), a_{1 k_1}) \overset{\alpha}{\longmapsto} \\[2mm] (\varphi_{h_2}(h'_{i_2 1}), a_2) \overset{\alpha}{\longmapsto} \end{cases}$$

$$(+) \left\{ \begin{array}{l}
(\varphi_{h_2}(h'_{i_2 2}), a_{22}) \xmapsto{\;\alpha\;} \ldots \xmapsto{\;\alpha\;} (\varphi_{h_2}(h'_{i_2 k_2}), a_{2k_2}) \xmapsto{\;\alpha\;} \\[2mm]
(\varphi_{h_3}(h'_{i_3 1}), a_3) \xmapsto{\;\alpha\;} \\[2mm]
(\varphi_{h_3}(h'_{i_3 2}), a_{32}) \xmapsto{\;\alpha\;} \ldots \qquad \ldots \qquad .
\end{array} \right.$$

Here $\varphi_{h_j}(h')$ denotes the isomorphic image of the half-edge h' in the hypervertex corresponding to h_j, under the isomorphism of this hypervertex onto the corresponding fragment F'_i in the right-hand sides of the rules of σ according to $(*)$.

Especially, if α searches, masters, or escapes from all labyrinths $\sigma(L)$, for $L \in \mathcal{L}$, then $\alpha^{(\sigma)}$ accomplishes this task for the labyrinths $L \in \mathcal{L}$. //

$\alpha^{(\sigma)}$ simulates the behaviour of α in a rather detailed manner. Often a "short form" $\alpha_0^{(\sigma)}$ of this automaton is sufficient or even more useful.

Let

$$\alpha_0^{(\sigma)} = (\overline{X}, \overline{Y}, \widetilde{A}, \widetilde{\delta}, \widetilde{\lambda}, \widetilde{a}_0),$$

where $\overline{X}$ and $\overline{Y}$ are defined as above, but

$$\widetilde{A} = \{\widetilde{a}_0\} \cup A.$$

We define

$$\widetilde{\delta}(\overline{x}, \widetilde{a}_0) = \widetilde{a}_0 \quad \text{and} \quad \widetilde{\lambda}(\overline{x}, \widetilde{a}_0) = \S$$

if the automaton α, starting on position $h'_{i(\overline{x}) 0}$ in the graphoid $F'_{i(\overline{x})}$, never crosses a free half-edge.

If α, starting on $h'_{i(\overline{x}) 0}$ in $F'_{i(\overline{x})}$, after finitely many steps leaves this graphoid with the state a' through the half-edge $h'_{i(\overline{x}), j_1}$, then let

$$\widetilde{\delta}(\overline{x}, \widetilde{a}_0) = a' \quad \text{and} \quad \widetilde{\lambda}(\overline{x}, \widetilde{a}_0) = \overline{y}(h_{i(\overline{x}), j_1}),$$

where $\overline{y}(h_{i(\overline{x}), j_1})$ is the output symbol causing the move from $h_{i(\overline{x}), j(\overline{x})}$ through $h_{i(\overline{x}), j_1}$ in the star $S_{i(\overline{x})}$.

Correspondingly, for any $a \in A$, let

$$\widetilde{\delta}(\overline{x}, a) = a \quad \text{and} \quad \widetilde{\lambda}(\overline{x}, a) = \S$$

if α, entering $F'_{i(\overline{x})}$ through the half-edge $h'_{i(\overline{x}), j(\overline{x})}$ in the state a, never leaves this graphoid; but

$$\widetilde{\delta}(\overline{x}, a) = a' \quad \text{and} \quad \widetilde{\lambda}(\overline{x}, a) = \overline{y}(h_{i(\overline{x}), j_1})$$

if, in this case, α leaves $F'_{i(\overline{x})}$ through $h'_{i(x), j_1}$ in the state a', and $\overline{y}(h_{i(\overline{x}),j_1})$ causes the move from $h_{i(\overline{x}),j(\overline{x})}$ through $h_{i(\overline{x}),j_1}$ in $S_{i(\overline{x})}$.

<u>Lemma 11.</u> Under the suppositions of Lemma 10 it holds

$\quad$ (+) iff

$$(h_1,a_0) \xmapsto{\alpha_0^{(\sigma)}} (h_2,a_2) \xmapsto{\alpha_0^{(\sigma)}} (h_3,a_3) \xmapsto{\alpha_0^{(\sigma)}} \ldots \quad \ldots \quad .$$

If α remains within some hypervertex of $\sigma(L)$ ad infinitum, then the behaviour of $\alpha_0^{(\sigma)}$ becomes stationary on the half-edge in L which corresponds to the free half-edge through which this hypervertex has been entered, and in the state of this last step in which the hypervertex is entered.
If α searches, masters, or escapes from all labyrinths $\sigma(L)$, for $L \in \mathcal{L}$, then $\alpha_0^{(\sigma)}$ does this task for all labyrinths from $\mathcal{L}$. //

Remark that the new initial state $\widetilde{a}_0$ is only introduced to enable the simulating automaton to start with an arbitrary position in the hypervertex corresponding to the starting position of α. If the simulation always starts with the image of the free half-edge $h_{i'(\overline{x}),j(\overline{x})}$ of $F_{i'(\overline{x})}$, we could take $\widetilde{a}_0 = a_0$; see Section 3.6 for an example.

<u>Hints & Sources.</u> Without detailed formal definitions, simulations by vertex substitutions for R-graphs are used by M. Blum and D. Kozen /L.BlKo/, other applications are given in /L.Da/ and /L.He82/. We are aware that the treatment in this section remains rather incomplete, too, but it seems to be sufficient to describe the basic idea. Later on this method, or even modifications of it, will be used for more comfortable types of automata, too.

<u>1.9.</u> <u>Edge substitutions</u>

In R-graphoids there is only one type of edges in a certain sense. Therefore, in this case an <u>edge substitution</u> σ consists of one rule,

$$\sigma = \{e \longrightarrow F'\},$$

where $F' = (L', h_1', h_2')$ is a <u>symmetric</u> R-corridor. The latter means that the corridor (L', h_2', h_1') is isomorphic to F'.

If L is an arbitrary R-graphoid with the edge set E_L, we assign to any $e \in E_L$ an isomorphic copy
$$F'_e = (L'_e , h'_{e1} , h'_{e2})$$
of F' such that L'_{e_1} and L'_{e_2} have no common elements, for $e_1 \neq e_2$.

Let
$$\sigma (L) = (\overline{V} , \overline{H} , \overline{I} , \overline{r})$$
with
$$\overline{V} = V_L \cup \bigcup_{e \in E_L} V_{L'_e} ,$$

$$\overline{H} = H_L \cup \bigcup_{e \in E_L} H_{L'_e} ,$$

$$\overline{r} = r_L \cup \bigcup_{e \in E_L} r_{L'_e} ,$$

$$\overline{I} = (I_L \setminus (H_L \times H_L)) \cup \bigcup_{e \in E_L} I_{L'_e} \cup \hat{I} ,$$

where
$$\hat{I} = \left\{ (h_i , h'_{ei}) , (h'_{ei} , h_i) : i = 1,2 \text{ and } e = \{ h_1 , h_2 \} \in E_L \right\}.$$

This part $\hat{I}$ of the incidence relation $\overline{I}$ causes the insertions of the R-corridors F'_e into the edges e.

From the symmetry of F' it follows that $\sigma(L)$ is uniquely determined by L up to isomorphism. Fig. 25 sketches an example.

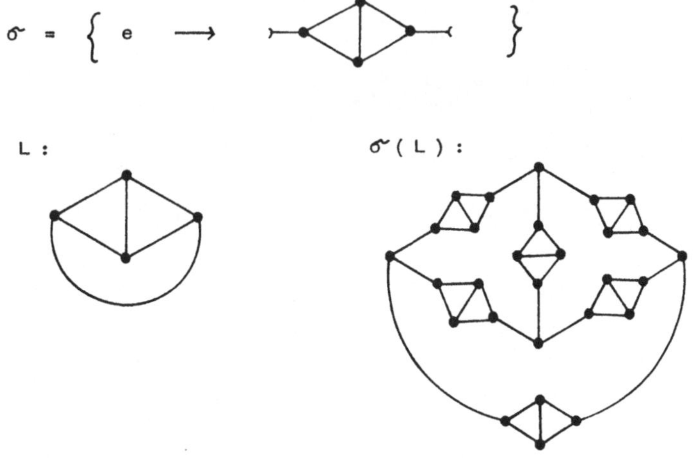

Fig. 25

In C-graphoids, edges can be distinguished each from the other by the directions of their half-edges. In the following we suppose that there is a permutation $\overline{}$ of the underlying direction set D such that

$$\bar{\bar{d}} = d \text{ ,for any direction } d \in D \text{ , and}$$

$$c_L(h_1) = \overline{c_L(h_2)} \text{ ,for any edge } e = \{ h_1 , h_2 \}$$

in a labyrinth L of the considered type $\mathcal{L}$. Intuitively,the mapping $^-$ gives the opposite direction,cf. Section 1.3 .

Let $D_0 \subseteq D$ be a fixed minimal set such that,for any direction $d \in D$, either d or $\bar{d}$ belongs to D_0.

An _edge_ _substitution_ on the type $\mathcal{L}$ of C-graphoids with the direction set D is a system of rules,

$$\sigma = \{ e_d \longrightarrow F_d' \; : \; d \in D_0 \} ,$$

where the F_d' are C-corridors of the form

$$F_d' = (L_d' , h_{d\,1}' , h_{d\,2}')$$

with $c_{L_d'}(h_{d\,1}') = d$ and $c_{L_d'}(h_{d\,2}') = \bar{d}$,for $d \in D_0$.

Given a labyrinth $L \in \mathcal{L}$,we assign to any element e of the edge set E_L an isomorphic copy

$$F_e' = (L_e' , h_{e\,1}' , h_{e\,2}')$$

of F_d' if $e = \{ h_1 , h_2 \}$ and $\{ c_L(h_1) , c_L(h_2) \} = \{ d , \bar{d} \}$.

Let L_{e_1}' and L_{e_2}' have no common elements if $e_1 \neq e_2$.

We define

$$\sigma (L) = (\bar{V} , \bar{H} , \bar{I} , \bar{c})$$

with

$$\bar{V} = V_L \;\cup\; \bigcup_{e \in E_L} V_{L_e'} \;\; ,$$

$$\bar{H} = H_L \;\cup\; \bigcup_{e \in E_L} H_{L_e'} \;\; ,$$

$$\bar{c} = c_L \;\cup\; \bigcup_{e \in E_L} c_{L_e'} \;\; ,$$

$$\bar{I} = (I_L \smallsetminus (H_L \times H_L)) \;\cup\; \bigcup_{e \in E_L} I_{L_e'} \;\cup\; \hat{I} \;\; ,$$

where

$$\hat{I} = \{ (h_i , h_{e\,i}') , (h_{e\,i}' , h_i) \; : \; i = 1,2 \text{ and } e = \{ h_1 , h_2 \} \in E_L \}.$$

By this definition,(the isoclass of) $\sigma(L)$ is uniquely determined by L. Fig. 26 shows an example for the type of 2D ficographs; let $D_{2,0} = \{ \text{north,west} \}$.

The graphoids L_e' inserted into the edges $e = \{ h_1 , h_2 \}$ of the labyrinth L together with the incident half-edges h_1 and h_2 are called the _hyperedges_ of $\sigma(L)$ in this context.

Now let σ be an edge substitution of the above described forms and α be a finite automaton able to work on the labyrinths $\sigma(L)$ for $L \in \mathcal{L}$. One easily defines a finite automaton $\alpha^{(\sigma)}$ which , working on a labyrinth $L \in \mathcal{L}$, simulates the behaviour of α on $\sigma(L)$, starting from a position $h \in H_L$.

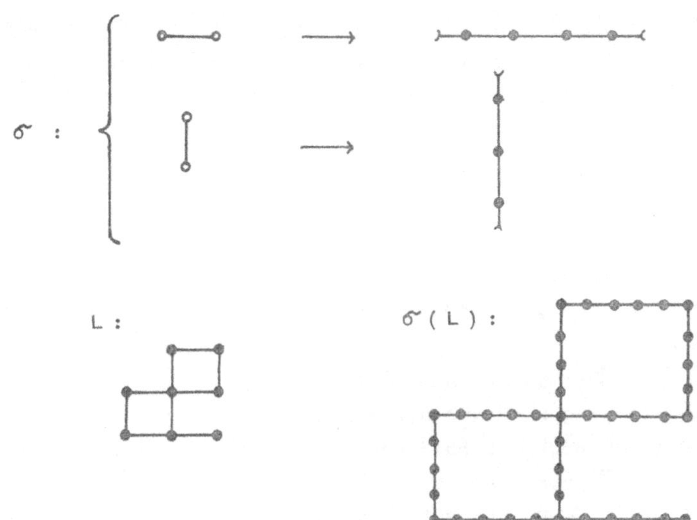

L :

$\sigma(L)$:

Fig. 26

If A is the state set of α, then let the state set of $\alpha^{(\sigma)}$ be equal to

$$\overline{A} = \begin{cases} A \cup (H_L. \times A) & \text{if } \mathcal{L} \text{ consists of R-graphoids,} \\ A \cup (\bigcup_{d \in D_0} H_{L_d}. \times A) & \text{if } \mathcal{L} \text{ consists of C-graphoids,} \end{cases}$$

where the $H_{L_d}.$ are assumed to be pairwise disjoint in the latter case.

To an α-configuration (h,a) with $h \in H_L$ on a labyrinth $\sigma(L)$, let correspond the $\alpha^{(\sigma)}$-configuration (h,a). If α is in a configuration (h_e' , a) with $h_e' \in H_{L_e.}$, for some edge $e = \{h_1, h_2\} \in E_L$, and h_1 is the last preceding α-position which belongs to H_L, then the corresponding $\alpha^{(\sigma)}$-configuration is given by $(h_1 , (h',a))$, where h' is the isomorphic image of h_e' in the hyperedge replacing e.

We omit the formal details of the definition of $\alpha^{(\sigma)}$ and consider only the short version $\alpha_0^{(\sigma)}$ in detail.

For $\alpha = (X,Y,A,\delta,\lambda,a_0)$ let $\alpha_0^{(\sigma)} = (X,Y,A,\overline{\delta},\overline{\lambda},a_0)$.

To any $x \in X$, let S_x denote the star corresponding to the input symbol x, and $h_x \in H_{S_x}$ be the position corresponding to x in S_x. The fragment $(S_x , h_1 , h_2 , \ldots , h_{1_x})$, where $h_1 = h_x$ and

$h_2, \ldots, h_{1_x}$ are the other free half-edges in the order given by
the rotation in $\mathrm{ver}(h_x)$, is uniquely determined up to isomorphism.

Moreover, let $F'_{h_i} = (L'_i, h'_{i1}, h'_{i2})$ be isomorphic copies of
the right-hand sides of the σ-rules corresponding to h_i such that
L'_i and S_x have pairwise no common elements ($1 \leqslant i \leqslant l_x$). Finally, let
$$\overline{F}_x = (\overline{L}_x, h'_{12}, \ldots, h'_{1_x 2})$$
denote the fragment obtained by interlinking the free half-edges
h_i of S_x with h'_{i1} in F'_{h_i} ($1 \leqslant i \leqslant l_x$). $\overline{F}_x$ is uniquely
determined by x up to isomorphism, too.

If the automaton α starting with the state $a \in A$ on position
$h_1 = h_x$ in the graphoid $\overline{L}_x$ never crosses a free half-edge, then
we define
$$\overline{\delta}(x,a) = a \quad \text{and} \quad \overline{\lambda}(x,a) = \S .$$
If α starting with the state a on position h_1 in $\overline{L}_x$ leaves this
graphoid through the half-edge h'_{j2} in the state a', then let
$$\overline{\delta}(x,a) = a' \quad \text{and} \quad \overline{\lambda}(x,a) = y(h_j) ,$$
where $y(h_j)$ be the output symbol causing the move from position
h_1 in S_x through the half-edge h_j ($1 \leqslant j \leqslant l_x$).

<u>Lemma 12.</u> Let $\beta = ((h_0,a_0), (h_1,a_1), \ldots)$ be the behaviour
    of a finite automaton α starting on some position $h_0 \in H_L$
    in a labyrinth $\sigma(L)$ with $L \in \mathcal{L}$,
    and β' denote the subsequence of β which consists of all
    configurations (h_i,a_i) with $h_i \in H_L$.

    If both β and β' are infinite, then β' is the behaviour of
    $\alpha_0^{(\sigma)}$ starting on position h_0 in the labyrinth L.

    If β is finite, this behaviour of $\alpha_0^{(\sigma)}$ is obtained from β'
    by adding a suitable pair (h',a') giving the half-edge and
    the state with which L is left.
    If β is infinite but β' is finite, the behaviour of $\alpha_0^{(\sigma)}$ is
    obtained from β' by adding an infinite part which simply
    repeats the last pair of β' infinitely often.
    Especially, if α searches, masters or escapes from all laby-
    rinths $\sigma(L)$ for $L \in \mathcal{L}$, then $\alpha_0^{(\sigma)}$ does this task for all
    labyrinths $L \in \mathcal{L}$.

We hope the reader accepts this lemma without a formal proof. //

<u>Lemma 13.</u> The problems to search all strongly normed and all
    normed 2D ficographs, respectively, by a finite C-automaton
    are equivalent.

If $f : \mathbb{N}^+ \longrightarrow \mathbb{N}^+$ such that,for any finite C-automaton α
with m states,there is a normed 2D trap (L,h) with at most
$f(m)$ vertices,then to any α there is also a strongly
normed 2D trap (L',h') with at most $2 \cdot f(m)$ vertices ,
and L' has the same number of faces as L .

Since the set of all strongly normed 2D ficographs is contained
in the set of all normed ones,for the equivalence we only have to
show that,given a finite C-automaton α searching all strongly
normed 2D ficographs,one can effectively construct an automaton
α' searching all normed 2D ficographs.

We consider the edge substitution

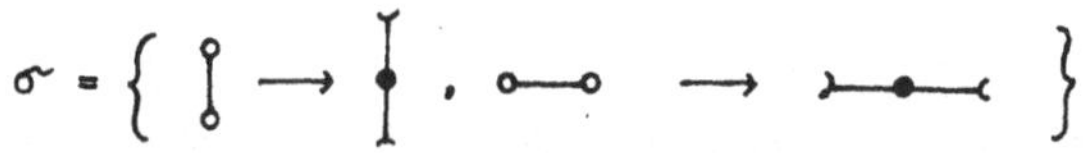

$$\sigma = \left\{ \; \substack{\circ \\ | \\ \circ} \longrightarrow \; \bullet \; , \; \circ\!\!-\!\!-\!\!\circ \; \longrightarrow \; \succ\!\!-\!\!\bullet\!\!-\!\!\prec \; \right\}$$

defined on the class of all normed 2D ficographs. For any normed
2D ficograph L,the labyrinth $\sigma(L)$ is strongly normed. Therefore,
since α searches all $\sigma(L)$, $\alpha' = \alpha_0^{(\sigma)}$ has the required property.

Conversely,if (L,h) is a normed 2D trap for $\alpha_0^{(\sigma)}$ with $\leqslant f(m)$
vertices,then ($\sigma(L)$, h) is a strongly normed 2D trap for α
with $\leqslant 2 \cdot f(m)$ vertices and as many faces as L has. Here let m be
the number of internal states of α and $\alpha_0^{(\sigma)}$. //

<u>Hints & Sources.</u> Although edge substitutions are very related to
vertex substitutions,they have not yet been explicitly investigated so far. Of course,Lemma 13 and the idea of its proof belong
to the folklore of labyrinth research. It analogously holds for
the mastering and the escaping problem (concerning open connected
2D graphoids in the latter case).

1.10. Edge insertions

Now we want to prove the equivalence of the searching problems
for the types of the normed 2D ficographs and the 2D ficographs,
respectively,by finite automata. Unfortunately,the 2D ficographs
cannot be transformed into normed ones by an edge substitution.
To this purpose,the more general concept of edge insertion will
be introduced. We shall consider it only on C-graphoids,the analogous construction on R-graphoids will be omitted here.
Moreover,as in the previous section,we presuppose the existence

of a mapping $^-$ giving the **opposite** direction such that $c_L(h_1) = \overline{c_L(h_2)}$ for any edge $\{h_1,h_2\}$ in a labyrinth L from the **considered** type $\mathcal{L}$. D_0 is a minimal set of directions such that either d or $\overline{d}$ belongs to D_0, for each direction $d \in D$.

Let L be a C-graphoid from $\mathcal{L}$ with the edge set E_L.

An <u>edge</u> <u>insertion</u> φ on L is based on a mapping assigning to any edge $e = \{h_1,h_2\}$ of a certain subset $E_\varphi \subseteq E_L$ a C-corridor

$$F'_{\varphi e} = (L'_e , h'_{e\,1} , h'_{e\,2})$$

such that

$$c_{L'_e} (h'_{e\,1}) \in D_0 \qquad \text{and}$$

$$\left\{ c_{L'_e} (h'_{e\,i}) : i = 1,2 \right\} = \left\{ c_L (h_i) : i = 1,2 \right\}.$$

We suppose that the corridors $F'_{\varphi\,e}$ and the labyrinth L have pairwise no common elements.

The C-graphoid obtained from L by the edge insertion φ is defined by

$$\varphi (L) = (\overline{V} , \overline{H} , \overline{I} , \overline{c})$$

with

$$\overline{V} = V_L \cup \bigcup_{e \in E_\varphi} V_{L'_e} \, ,$$

$$\overline{H} = H_L \cup \bigcup_{e \in E_\varphi} H_{L'_e} \, ,$$

$$\overline{c} = c_L \cup \bigcup_{e \in E_\varphi} c_{L'_e} \, ,$$

$$\overline{I} = (I_L \setminus \{ h: h \text{ is a half-edge of an edge from } E_\varphi \}^2)$$
$$\cup \bigcup_{e \in E_\varphi} I_{L'_e} \cup \hat{I} \, ,$$

where

$$\hat{I} = \left\{ (h_i,h'_{e\,i}) , (h'_{e\,i},h_i) : i = 1,2 \text{ and} \right.$$
$$\left. e = \{h_1,h_2\} \in E_\varphi , \ c_L(h_1) \in D_0 \right\}.$$

Let α be a finite C-automaton with the direction set D. If the sequence

$$\beta = ((h_0,a_0) , (h_1,a_1) , \ldots)$$

is a behaviour of α on $\varphi (L)$, then the subsequence

$$\beta_\varphi = ((h_{i_0},a_{i_0}) , (h_{i_1},a_{i_1}) , \ldots)$$

consisting of all elements (h_i,a_i) with $h_i \in H_L$ (those are the configurations on $\varphi(L)$ occupying positions from L) is called the <u>macrobehaviour</u> of α on $\varphi(L)$, corresponding to β.

Remark that β_φ can be finite, also if β is infinite. This is the case if α enters an **inserted** corridor $F'_{\varphi\,e}$ which it never leaves in the following steps.

By a <u>macrostep</u> of α ,we mean a sequence of consecutive steps of α ,starting with the i_j th one,for some element (h_{i_j},a_{i_j}) of β_φ , and terminating with the $(i_{j+1}-1)$ st step if $(h_{i_{j+1}},a_{i_{j+1}})$ is defined , or being infinite if (h_{i_j},a_{i_j}) is the last element of β_φ .

<u>Lemma 14.</u> Let φ_1 and φ_2 be edge insertions on a C-graphoid L such that $E_{\varphi_1} = E_{\varphi_2}$ and $F'_{\varphi_1\,e} \equiv_\alpha F'_{\varphi_2\,e}$ for all edges $e \in E_{\varphi_1}$ and some finite C-automaton α.
Then,for any starting position $h \in H_L$, the macrobehaviour of α on $\varphi_1(L)$ is equal to that of α on $\varphi_2(L)$. //

An edge insertion φ on a C-graphoid L with the direction set D_2 is said to be <u>rectilinear</u> if , for any edge $e \in E_\varphi$, the C-corridor $F'_{\varphi\,e}$ is a power of $\succ\!\!-\!\!\bullet\!\!-\!\!\prec$ or of $\begin{array}{c}\top\\\bullet\\\bot\end{array}$. This means,the C-corridors $F'_{\varphi\,e}$ have the forms shown in Fig. 27.

Fig. 27

Obviously,if φ is a rectilinear edge insertion and L is a 2D graphoid , then $\varphi(L)$ is a 2D graphoid,too , and it has the same number of faces as L .

<u>Lemma 15.</u> To any 2D ficograph L ,there is a rectilinear edge insertion φ such that $\varphi(L)$ is a normed 2D ficograph having at most n^2 vertices if n is the number of vertices of L .

To prove this,let P be a 2D embedding of L and

73

$$\delta^* = \min \left\{ |z-z'| : \begin{array}{l} \text{there are vertices } v,v' \text{ of } L \text{ such that} \\ z \text{ and } z' \text{ are either the x- or the y-} \\ \text{coordinates of } P(v) \text{ and } P(v'), \text{respect-} \\ \text{ively, and } z \neq z' \end{array} \right\}.$$

Then $\delta^* > 0$.

Now we consider a covering of the plane by a grid g of equidistant horizontal and vertical straight lines with the distance δ^* such that no embedding $P(v)$ of a vertex v lies on a line of g. It follows that any mesh of the grid g contains at most one embedding of a vertex.

Without loss of generality,we can suppose that any horizontal or vertical strip between two consecutive lines of the grid g contains a point $P(v)$,for some $v \in V_L$, if it contains a segment of an embedding $P(h)$ for $h \in H_L$. Indeed,as long as there exist strips of g which violate this condition,they can be removed by shifting all embeddings of vertices in suitable half-planes.

Now we define a rectilinear edge insertion φ on L.

Let $e \in E_L$. If e is a horizontal edge,let k_e be the number of vertical strips of g which are (completely) traversed by $P(e)$ ($= P(h_1) \cup P(h_2)$ if $e = \{h_1,h_2\}$). For a vertical edge e, let k_e be the number of horizontal strips traversed by $P(e)$. We define

$$E_\varphi = \left\{ e : e \in E_L \text{ and } k_e > 0 \right\},$$

$$F'_{\varphi e} = \begin{cases} (\;\rule[0.5ex]{0.6em}{0.4pt}\!\!\bullet\!\!\rule[0.5ex]{0.6em}{0.4pt}\;)^{k_e} & \text{if } e \text{ is a horizontal edge from } E_\varphi, \\[2ex] (\quad\rule[-0.8ex]{0.4pt}{2ex}\!\!\bullet\!\!\rule[-0.8ex]{0.4pt}{2ex}\quad)^{k_e} & \text{if } e \in E_\varphi \text{ is vertical} . \end{cases}$$

Fig. 28 shows an example how to obtain $\varphi(L)$ from L , namely by adding the crossed vertices.

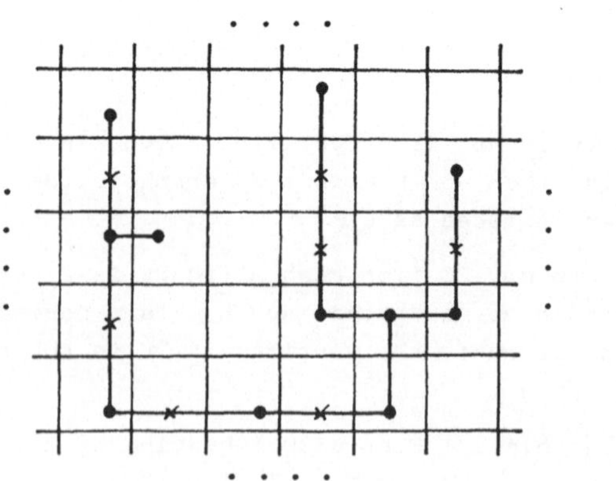

Fig. 28

Obviously, $\varphi(L)$ is a normed 2D ficograph.

By the additional property of the embedding P, both the number of vertical strips and that of horizontal strips which contain points from embeddings P(e) , for $e \in E_L$, are not greater than n , the number of vertices of L. Thus $\varphi(L)$ has at most n^2 vertices. //

<u>Lemma 16.</u> The problems to search all normed 2D ficographs and all 2D ficographs, respectively, by a finite automaton are equivalent.

Assume that $k \in \mathbb{N}^+$, $f : \mathbb{N}^+ \longrightarrow \mathbb{N}^+$, and to any finite automaton with m states, there is a 2D trap (L , h) with at most k faces and at most f(m) vertices. Then to any finite automaton having m states, there is a normed 2D trap (L', h') with at most k faces and at most $4 \cdot (\mu(m) \cdot f(m))^2$ vertices , where μ is the function from Proposition 3 .

To prove the first assertion, let α be a finite C-automaton searching all normed 2D ficographs and having m states. We consider the C-corridors

$$F_{ver} = \quad \text{and} \quad F_{hor} = \;\; .$$

By Proposition 3, there are numbers $k_{ver}, k_{hor} \leqslant \mu(m)$ such that

$$F_{ver}^{\,k_{ver}} \equiv_\alpha F_{ver}^{\,i \cdot k_{ver}} \quad \text{and} \quad F_{hor}^{\,k_{hor}} \equiv_\alpha F_{hor}^{\,i \cdot k_{hor}} .$$

for all $i \in \mathbb{N}^+$. For $k^* = k_{ver} \cdot k_{hor}$, we have

$$F_{ver}^{\,k^*} \equiv_\alpha F_{ver}^{\,i \cdot k^*} \quad \text{and} \quad F_{hor}^{\,k^*} \equiv_\alpha F_{hor}^{\,i \cdot k^*} .$$

We define an edge substitution σ by

$$\sigma = \left\{ \circ\!\!-\!\!\circ \;\rightarrow\; F_{hor}^{\,k^*} \;,\; \Big| \;\rightarrow\; F_{ver}^{\,k^*} \right\} .$$

For an arbitrary 2D ficograph L, let φ be a rectilinear edge insertion according to Lemma 15 , i.e., $\varphi(L)$ is normed and has $\leqslant n^2$ vertices if n gives the number of vertices of L.

Then $\sigma(\varphi(L))$ is also normed, hence it is searched by α. The 2D ficograph $\sigma(\varphi(L))$ can be considered to be obtained from L by an edge insertion which we simply denote by $\sigma \circ \varphi$. It inserts corridors of the form $F_{hor}^{\,i_e \cdot k^*}$ into horizontal edges e of L, and corridors $F_{ver}^{\,i_e \cdot k^*}$ into vertical edges e , where $i_e \in \mathbb{N}^+$.

By Lemma 14, α has the same macrobehaviour on $\sigma \circ \varphi(L)$ as on $\sigma(L)$, for any starting position from H_L. This means, α searches $\sigma(L)$, at least if it starts from positions in H_L. It follows that

$\alpha_0^{(\sigma)}$ searches L. The first assertion of Lemma 16 has been proved.

For the proof of the second assertion, let (L,h) be a 2D trap for $\alpha_0^{(\sigma)}$ with $\leq f(m)$ vertices, where m is the number of internal states of $\alpha_0^{(\sigma)}$ or, equivalently, of α. Let σ and φ be defined as above, for the given automaton α. Then $(\sigma \bullet \varphi(L), h)$ is a normed 2D trap for α, as the above considerations show.

$\sigma \bullet \varphi(L)$ has the same number of faces as L. Moreover, the number of vertices of $\varphi(L)$ is not greater than $(f(m))^2$, by Lemma 15. Therefore, $\varphi(L)$ has at most $2 \cdot (f(m))^2$ edges. It follows that the number of vertices in $\sigma \bullet \varphi(L)$ is bounded by

$$(k^* + 2) \cdot 2 \cdot (f(m))^2 \leq 2 \cdot k^* \cdot 2 (f(m))^2 \leq 4 \cdot (\mu(m) \cdot f(m))^2 . \quad //$$

<u>Hints & Sources.</u> The idea of replacing special corridors in labyrinths by α-equivalent ones is fundamental in L. Budach's trap construction /L.Bu75,78a,b/. So Lemma 14 and the first part of Lemma 16 are essentially known from Budach's papers. The second part of Lemma 16 involves H. Antelmann's /L.An/ size estimation of Budach's traps.

Über uns glänzten die Sterne,und die Visionen
der Bibliothek lagen weit in der Ferne.
"Wie schön ist die Welt und wie gräßlich sind
Labyrinthe ! " rief ich erleichtert aus.
"Wie schön wäre die Welt,wenn es eine Regel für
die Begehung von Labyrinthen gäbe!" sagte William.

Umberto Eco, Der Name der Rose

CHAPTER II. SEARCHING ALGORITHMS

In this chapter the fundamental searching algorithms for connected,especially for finite connected,R- and C-graphs are discussed. They are mainly given by programs written in Pidgin Programming Language (PPL). The algorithms are shown to be correct,
the types of automata able to implement them are characterized,
and the corresponding time complexities are estimated.

Throughout this chapter the number of vertices (that is the
size) of a finite labyrinth L is denoted by n_L . To avoid discussions of unessential special cases,we presuppose that any
labyrinth has at least two vertices. Thus,if it is connected,there
is an edge which is no loop. Remark,however,that this property
cannot be recognized by a finite R-automaton in the class of all
R-ficographs.

2.1. Pidgin Programming Language and Tarry's algorithm

Pidgin Programming Language (briefly PPL) is a slight modification of Pidgin ALGOL /S.AhHoUl/ and,like this,used to describe
algorithms in a clear and readable manner. It allows the use of
any notation and statement which have a defined or understandable
meaning in the corresponding context. Like Pidgin ALGOL , PPL has
no fixed set of data types and no explicite declarations of variables,but contains the traditional programming language constructs.

A PPL program for describing a searching algorithm consists of
an object specification followed by a (mostly compound) statement.
The object specification has the form

          <u>object</u> : specification ;

,where the specification characterizes the type of labyrinths on
which the program acts and the notations used for the components
of these objects. The aim of this construct is to improve the
readability of programs. Usually,the input of the described algorithm consists of a labyrinth of the specified type together
with a starting position in it.

    The PPL statements mainly used for our purposes are of the
following types.

    (1)   Denotation statement:
         <u>let</u> variable = expression
    (2)   Assignment statement:
         variable := expression
    (3)   Conditional statement:
         <u>if</u> condition <u>then</u> statement
       or
         <u>if</u> condition <u>then</u> statement <u>else</u> statement
    (4)   Iteration statement:
         <u>loop</u> statement ; ... statement <u>repeat</u>
       ,where the sequence of statements may contain the
       Exit statement:
         <u>exit</u>
    (5)   Compound statement:
         <u>begin</u> statement ; ... statement <u>end</u>
    (6)   Procedure call:
         procedure-name parameter-list
    (7)   Any miscellaneous statement having an understandable
       meaning.

    The meanings of statements of the types (2),(3),(5) and (7) are
assumed to be clear. A statement of type (1) declares a notation
for the thing described by the expression. It causes no action of
the implementing automaton.

    An iteration statement (4) creates a loop which has infinitely
often to be repeated if the sequence of statements between the
brackets <u>loop</u> and <u>repeat</u> does not contain a halt statement or an
exit statement. The exit statement,when it is reached,causes a
jump to the first statement following the corresponding keyword
<u>repeat</u>. The more known while - do and repeat - until statements can
easily be expressed by the iteration statement.

    By a procedure call (6),a procedure defined outside the program
can be activated. The parameter list may be empty. Mostly the
meaning of a procedure will be given by an informal description,

78

or it will be immediately understandable. For instance,we use the
predefined procedures

       stop

(halt statement) which causes the halt of the automaton,

       deposit

and

       pick up

concerning certain markers (pebbles or pointers), and

       move to

causing the move of the automaton to some new position.

As usual,too,calls of functions may occur in expressions. We
list some predefined functions together with their parameters,

       $H(v)$,
       $ver(h)$,
       $hal(h)$,
       $ang(h)$,
       $ord(v)$, $ord(h)$, $ord(e)$,
       $card(S)$.

Finally,we shall try to improve the readability of PPL programs
by comments. Comments are sentences included in the brackets (:
and :) . They are ignored by the automata implementing the programs.

As a first example of a PPL program and in order to illustrate
some further notations,we describe a modification of Tarry's
method for searching all ficographs.

Tarry's original idea was to traverse every edge of the graph
at most once in every direction,but to go through the entering
edge of a vertex only if there is no other possibility. The
entering edge is that edge incident to the vertex through which
it was entered at the first time.

The implementation of this idea would require to mark the directions in which the edges are already crossed. This could be
done by pointers. But,moreover,we would have to mark the entering
edge for every vertex,and this would require a second kind of
pointers. To avoid the use of two distinguishable types of
pointers,we have modified Tarry's original idea.

We use the pointers to mark the entering half-edge of a vertex
when it is reached at the first time. Therefore,at any point of
time,we can realize whether a certain vertex was already entered
at some earlier step. This is the case iff an incident half-edge
or a half-edge linked with an incident one carries a pointer. The
situation that no incident half-edge carries a pointer but a

half-edge linked with an incident one does carry a pointer is
possible only for the starting vertex.

From a position h incident to a vertex v we go forward through
another edge only if the corresponding neighbouring vertex has
not yet been entered in an actual step. To recognize these "free"
neighbours of v,an automaton has to perform some helping steps in
which,for all neighbours v' of v,it looks at the half-edges from
H(v') and at all half-edges linked with these. Then it enters the
first free neighbour in the sequence
$$ ver \circ hal \circ r(h) \,, ver \circ hal \circ r^2(h) \,, ver \circ hal \circ r^3(h) \,, \ldots \,, $$
where r denotes the rotation system of the given labyrinth.

If there are no free neighbours and the vertex v is not the
starting vertex (i.e.,there is an incident half-edge carrying a
pointer),then the automaton goes back through the entering edge;
otherwise (if v is the starting vertex) it halts.

This method would be successful on simple ficographs. But if
there is a loop incident to the starting vertex,it would be possible that the starting vertex is also entered in the first actual
step,and that loop becomes the entering edge. Then the above algorithm would become cyclic in going back from this vertex in
some later step.

To avoid this situation,we secure that in a first step a vertex
different from the starting vertex is entered. Now the starting
vertex can really be recognized in the later steps as the only
vertex having no entering edge but being the tail of the entering
edge for another vertex.

<u>Program 1.</u>

<u>object</u> : an R-ficograph $L = (V,H,I,r)$ of degree bound $b \geqslant 3$;
<u>begin</u>   <u>let</u> h = current position ; <u>let</u> v = ver(h) ;
          move to the first half-edge h' in the sequence
$$ (hal(h) \,, hal \circ r(h) \,, \ldots \,, hal \circ r^{deg(v)-1}(h)) $$
                   for which ver(h') $\neq$ v ;
          deposit a pointer on h' ;
      <u>loop</u> <u>let</u> h = current position ; <u>let</u> v = ver(h) ;
          N := $\{$ i : $1 \leqslant i < deg(v)$ and no half-edge from
                   H (ver $\circ$ hal $\circ$ r^i(h)) $\cup$ hal(H(ver $\circ$ hal $\circ$ r^i(h)))
                   carries a pointer
          (: indices of free neighbours of v with respect to h :) ;
          <u>if</u> N $\neq \emptyset$
          <u>then</u> <u>begin</u> <u>let</u> i_0 = min(N) ; <u>let</u> h_0 = r^{i_0}(h) ;
                   move to position hal(h_0) and
                   deposit a pointer here
               <u>end</u>

        <u>else</u> <u>if</u> there is an $h_1 \in H(v)$ which carries a pointer
        <u>then</u> move to position $\mathrm{hal}(h_1)$
        <u>else</u> stop
  <u>repeat</u>
<u>end</u>

Remember that the denotation statements (<u>let</u> ...) do not cause actions,they only introduce variables to denote certain things. Of course,sometimes the distinctions between denotation and assignment statements depend on the implementation.

In discussions of searching programs,a sequence of steps beginning with a statement of the form

     <u>let</u> variable = current position

and terminating either with a step followed by the next execution of such a denotation statement or with a halt statement,will be called a <u>macrostep</u> (of the program or of a performing automaton).

The above program starts with a macrostep corresponding to the program part from the first <u>begin</u> up to <u>loop</u>. After that,every execution of the sequence of statements within the loop corresponds to a further macrostep. Fig. 1 shows the situation and the hitherto occupied positions in a certain R-graph after 11 macrosteps of our program. The positions of pointers are marked by arrows.

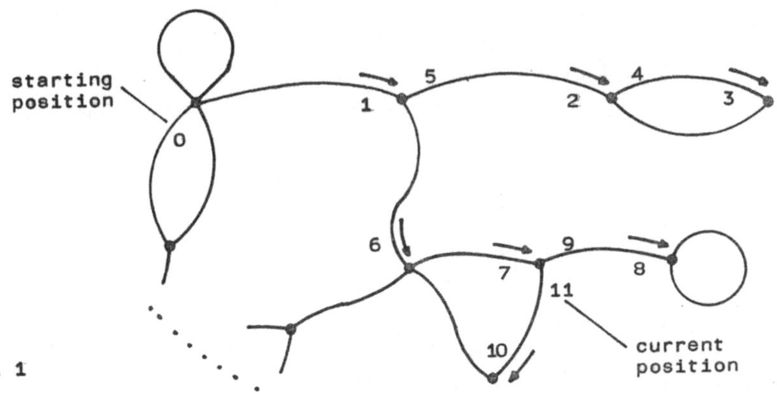

Fig. 1

Program 1 can be performed by a pointer automaton. In order to decide whether an edge $\{h_1, h_2\}$ with $h_1 \in H(v)$ is a loop , in the initial macrostep the automaton moves to position h_2, deposits a pointer there and goes back to h_1. Now by visiting all half-edges $h' \in H(v)$ (by moving to position $\mathrm{hal}(h')$ and going back to h')

the automaton recognizes whether the pointer is carried by such a
half-edge. $\{h_1,h_2\}$ is a loop iff this is the case for some
$h' \in H(v)$. After this recognition procedure, the pointer must be
picked up from h_2.

We see that the implementation of a macrostep (and even of a
PPL statement) may require a considerable number of elementary
steps of the performing automaton.

In an analogous way the set N and the half-edges h_0 or h_1 in
the following macrosteps can be obtained without the use of additional pointers.

<u>Theorem 1.</u> A pointer automaton working according to Program 1
searches every R-ficograph L of the corresponding degree
bound and halts after $O(n_L)$ steps.

A macrostep of Program 1 may be a forward macrostep yielding
the marking of a further half-edge by a pointer, namely if $N \neq \emptyset$.
If $N = \emptyset$ and the current vertex is not the starting one, we have a
backward macrostep. Finally, if $N = \emptyset$ at the starting vertex, the
halting macrostep is reached.

Let $G_L = (V,H,I)$ be the graph of the given labyrinth L, and
$G_k = (V_k,H_k,I_k)$ be the subgraph generated by all edges carrying
pointers after the k th macrostep and the incident vertices.
More precisely,
$$H_k = \left\{ h \in H : \text{ either } h \text{ or } hal(h) \text{ carries a pointer after} \atop \text{the k th macrostep} \right\},$$
$$V_k = \left\{ v \in V : \ H(v) \cap H_k \neq \emptyset \right\},$$
I_k is the restriction of relation I to $V_k \cup H_k$.

<u>Lemma 1.</u> Suppose that there is a k th macrostep. Then G_k is a
tree, and after the k th macrostep one obtains a path connecting an arbitrary vertex of G_k with the starting vertex
simply by going through the marked edges in the directions
opposite to the pointer directions.

This easily follows by induction on k. //
Since a pointer deposited after some macrostep is never picked
up in a later step and at most one of two incident half-edges can
carry a pointer, the lemma implies that every half-edge can be the
current position of at most one non-initial macrostep. Therefore,
after $O(n_L)$ macrosteps the halt statement must be executed, and
this can be only at the starting vertex. Since the number of elementary steps within any macrostep can be bounded by a constant,
we obtain the run time $O(n_L)$ for Program 1.

To prove Theorem 1 , it remains to show that every **vertex is the**
current vertex of some macrostep.

Otherwise it would exist a **vertex** v being the current vertex
of some macrostep but having a neighbour v' which never is the
current vertex. Obviously, v cannot be the starting vertex. Thus
it has an entering half-edge $h \in H(v)$. Let $h' \in H(v)$ be a half-edge
leading to v' , i.e. v' = ver (hal(h')), and
$$j_0 = \min \left\{ i : i \geqslant 1 \text{ and } r^i(h) = h' \right\} .$$
We consider the last macrostep whose current position belongs
to $S = \left\{ r^i(h) : 0 \leqslant i \leqslant j_0 \right\}$; let h'' be this position, see Fig. 2.

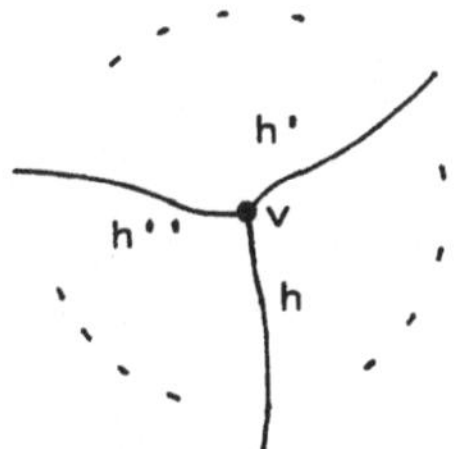

Fig. 2

If $1 \leqslant j' < \deg(v)$ and $r^{j'}(h'') = h'$, we have $j' \in N$ for this
macrostep. Therefore, it must exist an index $i_0 = \min (N) < j'$
in this macrostep. But then there is a later macrostep with the
current position $r^{i_0}(h'') \in S$. This contradiction completes the
proof. //

We see that Program 1 successively generates a spanning tree
of the given labyrinth L. If it halts after k macrosteps, then G_k
is a spanning tree of G_L .

Performing the initial macrostep of Program 1 , a pointer automaton must both deposit and pick up pointers. By a slight modification, one obtains an automaton which never picks up pointers.
For this purpose, let the automaton mark the starting vertex v by
pointers on all incident half-edges if $\deg(v) > 1$. If $\deg(v) = 1$,
no special marking of the starting vertex is needed.

<u>Hints & Sources.</u> Pidgin ALGOL was introduced in /S.AhHoUl/. For
basic principles of programming languages the reader is referred
to /S.Ho/. Tarry's searching method is completely presented by
D. König /L.Ko/.

2.2. Face-following and edge-blocking

Face-following procedures are essential parts of many labyrinth algorithms. Obviously, R-ficographs having only one face can be completely searched by following the face. The corresponding program is very simple.

Program 2a.

<u>object</u> : an R-ficograph $L = (V,H,I,r)$ of degree bound $b \nleq 3$
         with only one face ;
<u>begin</u> <u>loop</u> <u>let</u> h = current position ;
                 move to position $hal^\circ r(h)$
     <u>repeat</u>
<u>end</u>

This program can be performed by a finite R-automaton, but it never halts. In order to enable the automaton to halt some time after the labyrinth has been searched, one can use a pointer to mark the starting position or a pebble which marks the starting vertex. We give the programs for both cases.

Program 2b.

<u>object</u> : as in Program 2a ;
<u>begin</u>
          <u>let</u> h = current position ;
          deposit a pointer on h ;
          move to position $hal^\circ r(h)$;
      <u>loop</u> <u>let</u> h = current position ;
          <u>if</u> h carries a pointer <u>then</u> stop
              <u>else</u> move to position $hal^\circ r(h)$
     <u>repeat</u>
<u>end</u>

Program 2c.

<u>object</u> : as in Program 2a ;
<u>begin</u>   <u>let</u> h = current position ; z := deg(ver(h)) ;
          deposit the pebble on ver(h) ;
          move to position $hal^\circ r(h)$;
      <u>loop</u> <u>let</u> h = current position ;
          <u>if</u> ver(h) carries a pebble <u>then</u> z := z - 1 ;
          <u>if</u> z = 0 <u>then</u> stop
                  <u>else</u> move to position $hal^\circ r(h)$
     <u>repeat</u>
<u>end</u>

A macrostep of one of the programs 2a,b,c corresponds to one
elementary step of the implementing automata. In Programs 2b and
2c the halt statement is reached after $O(n_L)$ macrosteps.

Theorem 2. A finite R-automaton working according to Program 2a
    searches every R-ficograph of degree bound b with only one
    face. A 1-pointer automaton implementing Program 2b or a
    1-pebble automaton implementing Program 2c search every such
    labyrinth L and halt after $O(n_L)$ steps. //

As remarked in Section 1.2,any finite R-tree has only one face.
Therefore ,

Corollary 1. There are a finite R-automaton which searches every
    finite R-tree of a fixed degree bound , and a 1-pebble automa-
    ton searching every such R-tree L and halting after $O(n_L)$
    steps.

A finite 2-dimensional maze is said to be simply connected if
its order of connectivity is equal to 1. According to Section 1.6,
the searching problem for such mazes (by finite C-automata whose
way of working is defined in a straightforward manner) is reducible to the searching problem for finite 2D trees by finite C-automata. The latter problem is reducible to the searching problem
for finite R-trees of degree bound four by corresponding finite
R-automata. It follows

Corollary 2. There are a finite automaton searching every simply
    connected finite 2-dimensional maze , and a 1-pebble automaton
    searching every such a maze and halting after $O(n)$ steps,
    where n is the number of cells of the maze. //

Now we shall show that the number of pointers used by an automaton searching all R-ficographs can be linearly bounded by the
number of faces of the given object. To do this,we combine the
face-following procedure with the so-called edge-blocking method.
The edge-blocking method is based on a successive impoverish-
ment of the structure of the given labyrinth by removing edges.
Certain edges are blocked in the course of the working of the
searching automaton which simulates the behaviour of a suitable
searching procedure in the labyrinth obtained by removing all
blocked edges. For this purpose,the automaton must be able to
recognize the blocked edges. They may be marked in some special
way or may be distinguishable from the other by their environments
in the labyrinth.
Remember that the essence of the given modification of Tarry's

method is to construct a spanning tree of the labyrinth. In contrast to the edge-blocking method,this is done by a successive enrichment of the tree structure,beginning with the starting vertex only.

A first example of edge-blocking is the simulation of a searching algorithm for normed 2D ficographs with k faces on strongly normed 2D ficographs corresponding to finite 2-dimensional mazes with k barriers,see Section 1.6. Here all horizontal edges having southern neighbouring edges are thought to be blocked, and a searching algorithm working on the reduced labyrinth (obtained from the original one by removing all blocked edges) is simulated. The purpose of the blocking is to join the unessential faces which do not correspond to barriers of the maze with essential ones.

In the following we use the edge-blocking method to reduce the searching problem for general R-ficographs to that for R-ficographs with only one face. We start with a lemma which is of fundamental importance for all methods connected with the order of vertices or edges in labyrinths.

<u>Lemma 2.</u> To any degree bound b,there is a 1-pointer automaton which computes the order characteristic of its starting position in an R-ficograph. More precisely,starting on some position h in an R-ficograph L of degree bound b,the automaton halts after $O(n_L)$ steps on position h with a state giving cha(h). The pointer is only used to mark the starting position.

To do this,the automaton deposits the pointer on its starting position h. Then it traverses through all faces touching the vertex ver(h) and successively generates the decomposition cha(h). //

Remark that cha(h) immediately determines the orders ord(h) and ord$($ver$(h))$,cf. Section 1.2.

Let $L=(V,H,I,r)$ be an R-graph and B be a set of edges of L. Then $L^{/B/}$ denotes the R-graph obtained from L by removing all edges of B. More precisely,

$$L^{/B/} = (\ V_{L/B/}\ ,\ H_{L/B/}\ ,\ I_{L/B/}\ ,\ r_{L/B/}\ )\ ,$$

where

$$V_{L}/B/ = V\ ,$$

$$H_{L}/B/ = H \smallsetminus \left\{ h :\ \text{there is an edge}\ e \in B\ \text{with}\ h \in e \right\},$$

$$I_{L}/B/\ \text{is the restriction of I to}\ V \cup H_{L}/B/\ ,$$

and $r_L/B/$ is generated by r on $H_L/B/$, i.e.
$$r_L/B/ (h) = r^{i_h}(h)$$
if $i_h = \min \{i : i \in \mathbb{N}^+ \text{ and } r^i(h) \in H_L/B/\}$, for any $h \in H_L/B/$.

For the following program, an edge $e = \{h_1, h_2\}$ of a labyrinth is said to be <u>blocked</u> if it carries exactly one pointer, i.e., if exactly one of its half-edges h_1 and h_2 carries a pointer.

Moreover, to avoid misunderstandings, the functions cha and ord carry indices giving the denotations of the corresponding labyrinths.

Let h be a half-edge in some R-ficograph L_0 , and $v = \mathrm{ver}(h)$. If $\mathrm{ord}_{L_0}(v) > 1$, then obviously there is an edge e incident to v such that $h \not\in e$, $L_0/\{e\}/$ is an R-ficograph (i.e., it is connected) and $\mathrm{ord}_{L_0/\{e\}/}(v) = \mathrm{ord}_{L_0}(v) - 1$. Repeating this process, we finally obtain a set B_0 of edges incident to v such that $h \not\in e$ for all $e \in B_0$, $L_0/B_0/$ is an R-ficograph , and $\mathrm{ord}_{L_0/B_0/}(v) = 1$.

<u>Program 3.</u>
<u>object</u> : an R-ficograph $L = (V, H, I, r)$ of degree bound $b \geq 3$;
<u>begin</u> <u>let</u> $h =$ current position ; <u>let</u> $v = \mathrm{ver}(h)$;
        move to the first half-edge h' in the sequence
               $(\mathrm{hal}(h), \mathrm{hal} \circ r(h), \ldots, \mathrm{hal} \circ r^{\deg(v)-1}(h))$
               for which $\mathrm{ver}(h') \neq v$;
        <u>let</u> $h =$ current position ; <u>let</u> $v = \mathrm{ver}(h)$;
        compute $\mathrm{cha}_L(h)$;
        <u>if</u> $\mathrm{ord}_L(v) > 1$ <u>then</u>
            block some edges incident to v by depositing pointers
            on half-edges from $H(v) \setminus \{h\}$ such that the labyrinth
            remains connected but v becomes of order 1
            by removing all blocked edges ;
        deposit pointers both on h and on $\mathrm{hal}(h)$;
        move to position $\mathrm{hal}(h)$;
        $z := 0$ (: z counts the crossings of the specially marked
                                        basic edge :) ;
        <u>loop</u> <u>let</u> $h =$ current position ; <u>let</u> $v = \mathrm{ver}(h)$;
            <u>if</u> both h and $\mathrm{hal}(h)$ carry pointers <u>then</u>
                <u>begin</u> $z := z + 1$; <u>if</u> $z > 2$ <u>then</u> stop <u>end</u> ;
            <u>let</u> $B =$ set of all blocked edges of L ;
            compute $\mathrm{cha}_L/B/(h)$;
            <u>if</u> $\mathrm{ord}_L/B/(v) > 1$ <u>then</u>

$$\text{block some edges incident to } v \text{ by depositing}$$
pointers on half-edges from $H(v) \smallsetminus \{h\}$ such that
the labyrinth remains connected but v becomes
of order 1 by removing all blocked edges ;
>   <u>let</u> B = set of all edges blocked now ;
>   move to position $\mathrm{hal} \circ r_{L/B/}(h)$
>
>   <u>repeat</u>

<u>end</u>

In the first macrostep a pointer automaton performing Program 3
moves to an edge which is not a loop. This edge will be called
the <u>basic edge</u> in the sequel.

The second macrostep treats an endvertex of this basic edge by
blocking some other incident edges such that the order of the
vertex becomes 1 but the labyrinth remains connected after removing the blocked edges. This is possible by our preparatory remark
and can be done in some well-defined way which the reader may
specify.

Now the basic edge is marked by two pointers,the counter variable z is initialized,and the automaton moves to the other end-
point of the basic edge.

From now on the face-following in the labyrinth obtained by
removing all blocked edges is simulated. But at any reached vertex
having an order >1 in this labyrinth some incident edges are
blocked in a suitable well-defined manner such that the vertex
obtains the order 1. By this way,any non-basic edge is marked by
at most one pointer,and the basic edge can uniquely be recognized
at any time.

To complete the program,we have to explain how the computation
of $\mathrm{cha}_{L/B/}(h)$ in the macrostep following the second one can be
performed. The problem is that in this procedure the half-edge h
cannot be marked by a pointer as sketched in the proof of Lemma 2.
Indeed,this marking could not be distinguished from the markings
of edges blocked by pointers.

Let h be the current position and $v = \mathrm{ver}(h)$,for some macrostep following the second one. Then the edge $\{h, \mathrm{hal}(h)\}$ cannot
be blocked and cannot be a loop. If $\deg(v) = 1$,we have $\mathrm{cha}(h) = \{\{0\}\}$.
Now let $\deg(v) > 1$.

If there is a half-edge from $H(v) \smallsetminus \{h\}$ which carries a pointer,
h must have been the current position in some earlier macrostep
following the second one,and we have $\mathrm{cha}_{L/B/}(h) = \{\{0\}\}$,too.

If no half-edge from $H(v) \smallsetminus \{h\}$ carries a pointer and there is

a half-edge h'' $\in$ H(ver∘hal(h)) such that hal(h'') $\neq$ h and
hal(h'') does not carry a pointer,position h can be uniquely
marked by pointers on all half-edges hal(h') , for h' $\in$ H(v) ,and
on h , see Fig. 3a.

If no half-edge from H(v)∖{h} but all hal(h'') $\neq$ h , for
h'' $\in$ H (ver∘hal(h)), carry pointers,position h can be marked by
pointers on h and hal(h) , see Fig. 3b .

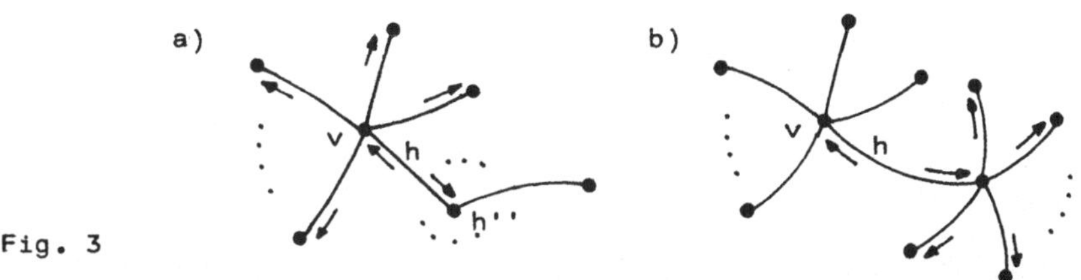

Fig. 3

In traversing all faces touching the vertex v = ver(h) in $L^{/B/}$
in order to compute $cha_{L/B/}(h)$, the automaton,when it meets an
edge carrying a pointer,has to decide whether this pointer belongs
to the marking of the current position h of the macrostep in the
way sketched above or whether it marks the basic edge or a blocked
edge. This decision can be done by visiting all edges incident to
the endvertices of the corresponding edge.

Now we have sufficiently specified how a pointer automaton has
to perform the given program.

<u>Theorem 3.</u> A pointer automaton working according to Program 3
searches every R-ficograph L of degree bound b , halts after
$O(n_L^2)$ steps and uses at most f + b + 2 = O (f) pointers
if L has exactly f faces .

The automaton needs two pointers to mark the basic edge. To
mark the current position of some macrostep in the course of the
computation of its characteristic,it needs at most b + 1 further
pointers. Finally,by every blocked edge two faces of the original
labyrinth L are joined. Hence the automaton can block at most
f - 1 edges. This leads to the given bound of the number of used
pointers.

If a vertex v is the current vertex of some non-initial macrostep,no further edge incident to v will be blocked in a later
macrostep. Therefore,the current position of the next macrostep
can be regarded to be obtained by a face-following step in the
labyrinth derived from L by removing all edges which are blocked
during an arbitrary macrostep. Let $\overline{B}$ denote this set of all edges

blocked in some macrostep. The above fact means that the sequence
of consecutive current positions corresponds to a face-following
path in the labyrinth $L/\overline{B}/$. Therefore,this sequence becomes
cyclic after $O(n_L)$ macrosteps.

By Lemma 1.2 , $L/\overline{B}/$ has only one face. Thus the automaton
searches the labyrinth $L/\overline{B}/$ and also L which has the same vertex set. But in the third macrostep whose current position belongs
to the basic edge the program halts. This is the case after $O(n_L)$
macrosteps,too.

Finally,we observe that every macrostep consists of at most
$O(n_L)$ elementary steps of the automaton,where the computation of
the order characteristic of the current position is the most time
consuming part. //

<u>Hints & Sources.</u> The statements of Theorem 2 and Corollaries 1
and 2 have been widely known for years already. The first part
of Corollary 2 was directly proved by K. Döpp /L.Do/. Although
the combination of face-following and edge-blocking is rather
simple,the searching result given by Theorem 3 has not yet been
published so far. It could be also proved by a simulation of
Tarry's algorithm on the dual of the given R-ficograph if this
dual R-ficograph is defined in a suitable manner.

The combination of edge-blocking with another fundamental
searching method,namely bridge-preferring,will be considered in
Section 2.4 . In Section 2.5 we shall give a further application
of the face-following method combined with edge-blocking which
works in normed 2D ficographs.

2.3. The bridge-preferring method

The original version of this method can be used to search plane
R-ficographs having no vertex of an order greater than two. Remember that,by Lemma 1.3,the bridges in plane R-graphs can simply be
recognized by computing the orders of edges.

The basic idea of <u>bridge-preferring</u> is as follows.

Entering a vertex of order 1,one has to follow the face.

At vertices of order 2 in plane labyrinths,there are exactly
two incident half-edges of order 2,say h_0 and h_0' . All the
other incident half-edges belong to bridges. Fig. 4 illustrates
this situation.

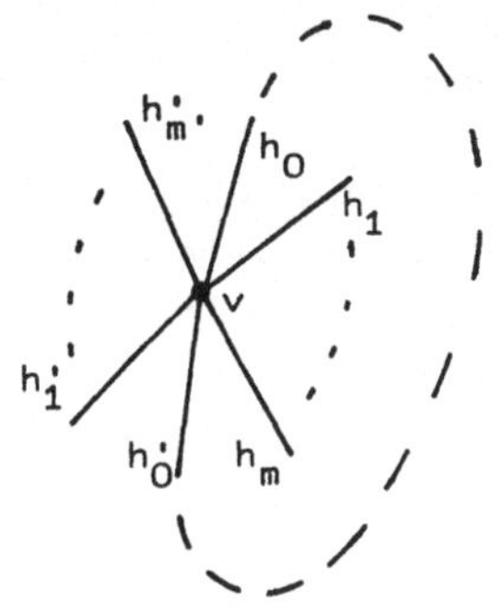

Fig. 4

Arriving at such a vertex through the half-edge h_0, one has to treat all incident bridges according to the succession $h_1,\ldots,h_m$, $h_1',\ldots,h_{m'}'$, and only after that one can cross through the half-edge h_0' . To treat a bridge means to search the whole shore of the corresponding neighbouring vertex in the labyrinth. To accomplish this idea, an automaton has only to store the information on which side (i.e. right or left) of the path (h_0,v,h_0') it is working in a certain stage.

Correspondingly, for any half-edge h of order 1 and incident to v, let $sid_{h_0}(h)$ denote the side on which it meets the path (h_0,v,h_0'). More precisely,

$$sid_{h_0}(h) = \begin{cases} \text{left} & \text{if } h = h_j \text{ for } 1 \leq j \leq m, \\ \text{right} & \text{if } h = h_j' \text{ for } 1 \leq j \leq m'. \end{cases}$$

The successor of a half-edge $h \in H(v) \smallsetminus \{h_0'\}$ with respect to the linear ordering sketched in Fig. 4 is denoted by $suc_{h_0}(h)$. Thus,

$$suc_{h_0}(h) = \begin{cases} r(h) & \text{if } ord(r(h)) = 1 , \\ r^2(h) & \text{if } r(h) = h_0' \text{ and } r^2(h) \neq h_0 , \\ h_0' & \text{otherwise} , \end{cases}$$

where r denotes the rotation system of the labyrinth. $suc_{h_0}(h_0')$ is undefined.

Obviously, if one knows $cha(h)$ and $sid_{h_0}(h)$ for some half-edge h of order 1 and incident to v, both $suc_{h_0}(h)$ and , if $ord(suc_{h_0}(h)) = 1$, $sid_{h_0}(suc_{h_0}(h))$ can be uniquely determined.

In the following program , the parameter s_0 can be an arbitrary element from $\{\text{left},\text{right}\}$.

<u>Program 4.</u>   (: Let $s_0 \in \{\text{left},\text{right}\}$. :)

<u>object</u> : a plane R-ficograph $L = (V,H,I,r)$ of degree bound $b \geq 3$
with no vertex of an order > 2 ;

```
begin  s := s_0 ;
    loop  let  h = current position ;  let  v = ver( h ) ;
          compute  cha( h ) ;
          if  deg(v) ≤ 2  or  ord(v) = 1
              then  move to position  hal( r(h) )
              else  (: deg(v) ≥ 3  and  ord(v) ≥ 2  :)
                    if  ord(v) = 2  then
                        begin  if  ord(h) = 2
                                   then  let  h_0 = h
                                   else  (: h belongs to a bridge :)
                                         let  h_0 ∈ H(v)  such that
                                              ord( h_0 ) = 2  and
                                              sid_{h_0}( h ) = s ;
                               move to position  hal∘suc_{h_0}( h ) ;
                               if  ord( suc_{h_0}(h) ) = 1
                                   then  s := sid_{h_0}( suc_{h_0}(h) )
                        end
    repeat
end
```

This program can be performed by a 1-pointer automaton if the
computation of cha(h) is executed according to Lemma 2. Then
any macrostep consists of $O(n_L)$ elementary steps of the
automaton.

Remark that the condition "if ord(v) = 2 then" could be omitted
because of our object specification. It is used in order to make
possible an easy modification of the program for more general
purposes.

Let h be a half-edge belonging to a bridge in some plane
R-ficograph $L = (V,H,I,r)$. By L_h we denote the shore containing
ver(h) together with this bridge $\{h, hal(h)\}$. More precisely, L_h
is the component that contains h of the R-subgraphoid generated
by the half-edges from
$$(H \smallsetminus H(ver∘hal(h))) ∪ \{ hal(h) \}$$
and the incident vertices. Fig. 5 shows an example. Remark that
a bridge of L cannot be a loop.

Lemma 3. Let a 1-pointer automaton work according to Program 4
     in a plane R-ficograph L, and h be a half-edge belonging to
     a bridge of L. Suppose that h is the current position of
     the k th macrostep and $s_k ∈ \{left, right\}$ is the value of the
     variable s in the result of this macrostep.

92

i) If L_h does not contain a vertex of an order $\geqslant 3$, then
there is a $k' > k$ such that $hal(h)$ is the current position of the k'th macrostep. If k' is the minimal
number $> k$ of this kind, the value of variable s at the
begin of the k'th macrostep is equal to s_k.
Moreover, to every half-edge h'' in L_h, there is a number k'' such that $k \leqslant k'' \leqslant k'$ and h'' or $hal(h'')$
is the current position of the k''th macrostep.

ii) If there is a vertex of an order $\geqslant 3$ in L_h, then the automaton enters such a vertex as the current vertex of
the k'th macrostep, for some $k' \geqslant k$, and it does not
change its current position after that.

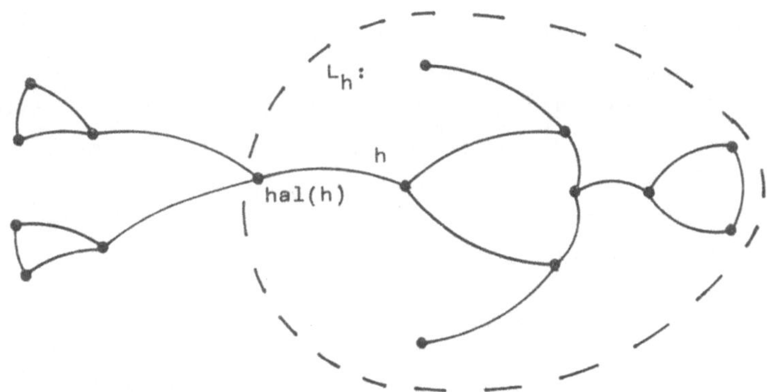

Fig. 5

To prove Lemma 3, we first remark that $cha_L(h') = cha_{L_h}(h')$ for
any half-edge $h' \neq hal(h)$ in the labyrinth L_h. This follows since
h belongs to a bridge and L is plane. Hence the macrobehaviour of
the automaton in L_h (i.e., the sequence of current positions and
states at the beginnings of the macrosteps) only depends on L_h.

Now we prove the lemma by induction on the number of faces of
the plane R-ficograph L_h.

If L_h has only one face, it is an R-tree. Then the assertion
of the lemma obviously holds.

Assume that the assertion holds for any position h in a plane
R-ficograph L, where L_h has at most m faces $(m \geqslant 1)$. Let L_h have
$m+1$ faces now.

We consider the (macro-) behaviour of the automaton beginning
with the kth macrostep in the current position h. If follows the
face containing $ang(h)$ in L_h as long as the current vertices
have the order 1. The value of the variable s is not changed in

this stage. Since there exist at least two faces in L_h , by
Lemma 2.2 , there is a macrostep whose current vertex has an order
$\geqslant 2$. Let h^* be the current position of the first macrostep of
this kind , and $v^* = ver(h^*)$ be the current vertex. Obviously ,
$ord(h^*) = 1$. Let $h_0 \in H(v^*)$ such that $ord(h_0) = 2$ and
$sid_{h_0}(h^*) = s_k$ (i.e., the current value of the variable s).

If $ord(v^*) > 2$, the assertion of Lemma 3 has been proved.

Now let $ord(v^*) = 2$.

There is a maximal simple path
$$w = (h_0, v_0, h_0', h_1, v_1, h_1', \ldots , h_l')$$
starting with h_0 and $v_0 = v^*$ such that $l \in \mathbb{N}$ and
$$ord(h_i) = ord(v_i) = ord(h_i') = 2 \quad \text{for } 0 \leqslant i \leqslant l .$$
Since w is maximal, we have either $ord(ver \circ hal(h_L')) > 2$ or
$hal(h_L') = h_0$.

If $0 \leqslant i \leqslant l$ and $\overline{h} \in H(v_i) \smallsetminus \{ h_i, h_i' \}$, the edge $\{ \overline{h}, hal(\overline{h}) \}$ is
a bridge in L_h , and for the half-edge $hal(\overline{h})$ the assertion of
the lemma is satisfied, since $L_{hal(\overline{h})}$ has at most m faces.
See Fig. 6 for illustration and remember that the faces correspond
to the regions.

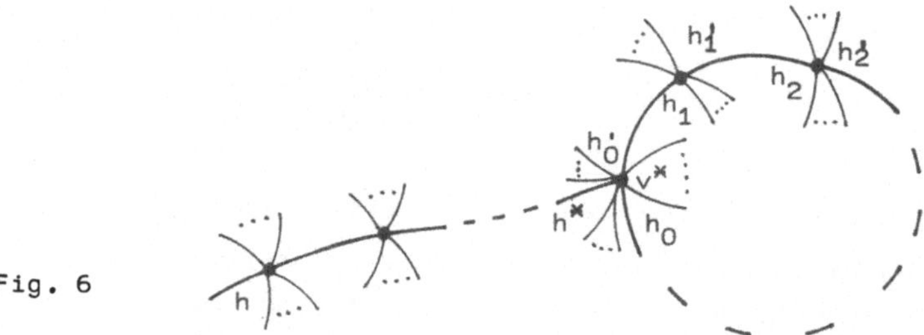

Fig. 6

If one of these labyrinths $L_{hal(\overline{h})}$ contains a vertex of an
order $\geqslant 3$ or if $ord(ver \circ hal(h_i')) \geqslant 3$, then the automaton
reaches a vertex of an order $\geqslant 3$ as the current vertex of some
macrostep. Otherwise, it finally reaches the current position
$hal(h_i') = h_0$ and it also enters the current position $hal(h^*)$ in
some later macrostep , but only after it has treated all bridges
incident to vertex v^*. Moreover, when $hal(h^*)$ is entered, the
variable s has the value s_k.

For all bridges incident to vertices in the simple path connecting position h^* with h, the assertion of the lemma applies by
the hypothesis of induction. Therefore, if L_h contains a vertex of
an order $\geqslant 3$, it is reached as the current vertex of some macro-

step. Otherwise,the automaton returns through the edge $\{h,hal(h)\}$ with the value s_k of the variable s. //

<u>Theorem 4.</u> A 1-pointer automaton that performs Program 4 searches every plane R-ficograph of the corresponding degree bound if it has no vertex of an order $\geqslant 3$.

This easily follows from Lemma 3. If the given plane R-ficograph does not contain a bridge,it must be a cycle,where every vertex has degree 2. Then the assertion trivially holds. Otherwise,the automaton reaches a half-edge of order 1 as the current position of some macrostep. Then the assertion follows from Lemma 3. //

<u>Corollary 3.</u> There is a 3-pointer automaton which searches every plane R-ficograph L having no vertex of an order $\geqslant 3$ and halts after $O(n_L^2)$ steps.

Indeed,the two additional pointers can be used to mark a basic edge analogously to the previous section , and the automaton can halt when it repeatedly crosses this basic edge. This is the case after $O(n_L)$ macrosteps each consisting of $O(n_L)$ elementary steps of the automaton. //

<u>Hints & Sources.</u> The basic idea of bridge-preferring was first sketched in /L.He84/ and /L.HeKr/,and in detail considered in /L.He86a/ , but always for the degree bound three.

2.4. Modifications of bridge-preferring

In this section we combine the bridge-preferring method with edge-blocking and trying all possible ways,respectively.

Like in Section 2.2,the edge-blocking method is used to reduce the number of faces. The aim is to eliminate all vertices of orders $\geqslant 3$. This is based on the following fact.

<u>Lemma 4.</u> In any plane R-ficograph possessing a vertex of an order $\geqslant 3$, a 1-pointer automaton working according to Program 4 reaches such a vertex as current vertex of some macrostep.

This easily follows by applying Lemma 3 in a similar way as in the proof of Theorem 4. //

As in Section 2.2,an edge is said to be blocked if it carries
exactly one pointer. By pointers on both half-edges,we mark a
special basic edge in order to obtain an always halting algorithm.
Also the notation $L^{/B/}$ and the related ones are used like in
Section 2.2 .

<u>Program 5.</u>   (: Let $s_0 \in \{\text{left},\text{right}\}$. :)

<u>object</u> : a plane R-ficograph $L = (V,H,I,r)$ of degree bound $b \geqslant 3$;

<u>begin</u> $s := s_0$;

    <u>let</u> h = current position ; <u>let</u> v = ver(h) ;

    $i_0 := \min \{ i : 0 \leqslant i < \deg(v)$ and $\text{ver} \circ \text{hal} \circ r^i(h) \neq v \}$;

    move to position $\text{hal} \circ r^{i_0}(h)$;

    <u>let</u> h = current position ; <u>let</u> v = ver(h) ;

    compute $\text{cha}_L(h)$;

    <u>if</u> $\text{ord}_L(v) \geqslant 3$

        <u>then</u>   block some edges incident to v by depositing
                 pointers on half-edges from $H(v) \setminus \{ h \}$ such that
                 the labyrinth remains connected but v becomes of
                 order 2 by removing all blocked edges ;

    deposit pointers on both h and hal(h)
                  (: this marks the basic edge :) ;

    move to position hal(h) ;

    $z := 0$ (: z counts the crossings of the basic edge
                     after the last blocking step :) ;

    <u>loop</u> <u>let</u> h = current position ; <u>let</u> v = ver(h) ;

        <u>if</u> both h and hal(h) carry pointers <u>then</u>
             <u>begin</u> $z := z + 1$; <u>if</u> $z > 2$ <u>then</u> stop <u>end</u> ;

        <u>let</u> B = set of all blocked edges of L ;

        compute $\text{cha}_{L/B/}(h)$;

        <u>if</u> $\deg_{L/B/}(v) \leqslant 2$ or $\text{ord}_{L/B/} = 1$

            <u>then</u>   move to position $\text{hal} \circ r_{L/B/}(h)$

            <u>else</u>   (: $\deg_{L/B/}(v) > 2$ and $\text{ord}_{L/B/}(v) > 1$:)

                 <u>begin</u> <u>if</u> $\text{ord}_{L/B/}(v) > 2$ <u>then</u>

                       <u>begin</u> block some edges incident to v
                       by pointers on half-edges from
                       $H(v) \setminus \{ h \}$ such that the laby-
                       rinth remains connected but v
                       is of order 2 after removing
                       all blocked edges ; $z := 0$;
                       <u>let</u> B = set of all blocked
                           edges of L
               <u>end</u> ;

$$\textbf{if} \quad \text{ord}_{L^{/B/}} (h) = 2$$

$$\underline{\textbf{then}} \quad \underline{\textbf{let}} \quad h_0 = h$$

$$\underline{\textbf{else}} \quad (: h \text{ belongs to a bridge of } L^{/B/} \text{ :})$$

$$\underline{\textbf{let}} \quad h_0 \in H(v) \quad \text{such that}$$

$$\text{ord}_{L^{/B/}} (h_0) = 2 \quad \text{and}$$

$$\text{sid}_{L^{/B/}, h_0} (h) = s \text{ ;}$$

$$\text{move to position } \text{hal} (\text{suc}_{L^{/B/}, h_0} (h)) \text{ ;}$$

$$\textbf{if} \quad \text{ord}_{L^{/B/}} (\text{suc}_{L^{/B/}, h_0} (h)) = 1$$

$$\underline{\textbf{then}} \quad s := \text{sid}_{L^{/B/}, h_0} (\text{suc}_{L^{/B/}, h_0} (h))$$

$$\underline{\textbf{end}}$$

$$\underline{\textbf{repeat}}$$

$$\underline{\textbf{end}}$$

Without going into details we remark that the computation of $\text{cha}_{L^{/B/}}(h)$ can be performed similarly to Program 3. Therefore, Program 5 can be accomplished by a pointer automaton.

By Lemma 4, if there is a vertex of an order $\geqslant 3$ in the current labyrinth $L^{/B/}$, the automaton always reaches such one in at most $O(n_L)$ macrosteps after starting from the basic edge. By blocking some edges, this vertex obtains the order 2, and the variable z is reset to 0 then. Therefore, if L contains only k vertices of orders $\geqslant 3$, after $O(k \cdot n_L)$ macrosteps a labyrinth $L^{/B/}$ without vertices of orders $\geqslant 3$ is obtained, and $O(n_L)$ macrosteps later the algorithm halts.

$\underline{\text{Theorem 5.}}$ Program 5 can be performed by a pointer automaton.

This searches every plane R-ficograph L of degree bound b, halts after $O((k+1) \cdot n_L^2)$ steps and uses at most $(b-2) \cdot k + b + 1 = O(k)$ pointers if L contains only k vertices of orders $\geqslant 3$. //

Now we combine the bridge-preferring method of Program 4 with a move management by control sequences. In order to obtain a clear representation, first we consider graphs of degree bound three. Later on it will be shown how to apply our idea to labyrinths of higher degrees.

In labyrinths of degree bound three, every half-edge incident to a vertex of order 3 has the order 2. For such current positions, the move of the automaton is controlled by a corresponding letter of the given control sequence $SEQ \in \{0,1,2\}^{\times}$. At current vertices of order 2, the following program behaves like Program 4.

With respect to current vertices of order 1,we have to distinguish between two possible stages of our program. These depend on
the direction,forward or backward,of the face-following at such
vertices. The corresponding state of the program is determined by
the variable f ,which changes its value at vertices of order 3 if
the current letter of the control sequence is equal to O.

__Program 6.__   (: Let SEQ $\in \{0,1,2\}^{\ast}$, $s_0 \in \{$left,right$\}$, $f_0 \in \{1,-1\}$:)
__object__ : a plane R-ficograph L = (V,H,I,r) of degree bound 3 ;
__begin__ f := f_0 ; s := s_0 ;
        z := 1 ;
        __loop__ __let__ h = current position ; __let__ v = ver(h) ;
              compute cha (h) ;
              __if__ deg(v) $\leqslant$ 2
                  __then__   move to position hal∘r (h)
                  __else__ (: deg(v) = 3   :)
                        __begin__ __if__ ord(v) = 1
                              __then__ move to position hal∘r^f(h) ;
                           __if__ ord(v) = 2 __then__
                              __begin__ __if__ ord(h) = 2
                                    __then__ __let__ h_0 = h
                                    __else__ __let__ $h_0 \in$ H(v) such
                                          that ord (h_0) = 2
                                          and sid_{h_0} (h) = s ;
                                 move to position
                                       hal∘suc_{h_0}(h) ;
                                 __if__ ord (suc_{h_0} (h)) = 1
                                       __then__   s := sid_{h_0} (suc_{h_0} (h))
                              __end__ ;
                        __if__ ord(v) = 3
                           __then__ __if__ z > len(SEQ)
                                 __then__   stop
                                 __else__ (: z $\leqslant$ len(SEQ) :)
                                 __begin__ __let__ i_0 = z th letter
                                       of SEQ ; z := z + 1 ;
                                    move to position
                                          hal∘r^{i_0}(h) ;
                                    __if__ i_0 = O __then__ f := -1·f
                              __end__
              __end__
                  __end__
        __repeat__
__end__

98

For $s \in \{\text{left},\text{right}\}$, let $\bar{s}$ denote the opposite direction, i.e. $\overline{\text{left}} = \text{right}$ and $\overline{\text{right}} = \text{left}$.

Lemma 5. Let $(h_1, h_2, \ldots, h_l)$ be a sequence of current positions of l consecutive macrosteps of Program 6, where
$\text{ord}(\text{ver}(h_i)) \leq 2$ for $1 \leq i < l$, $f = 1$ during all these macrosteps, and the sequence $(s_0, s_1, \ldots, s_l)$ gives the values
of the variable s; more precisely, s_{i-1} is the value of s at
the beginning, but s_i that at the end of the i th macrostep in
the sequence.
Then $(\text{hal}(h_l), \ldots, \text{hal}(h_2), \text{hal}(h_1))$ becomes the sequence
of current positions of l consecutive macrosteps starting
with $f = -1$ and $s = \overline{s_l}$ on the position $\text{hal}(h_l)$, and
$(\overline{s_l}, \ldots, \overline{s_1}, \overline{s_0})$ is the corresponding sequence of values
of s.

This can easily be shown by induction on l. //

By Lemmas 4 and 5, Program 6 can never work infinitely long
without reaching a macrostep whose current vertex has the order 3.

In the following, a vertex of order i is called an <u>Oi-vertex</u>,
and a macrostep whose current vertex is an Oi-vertex is called an
<u>Oi-macrostep</u> $(i = 1,2,3)$.

For a sequence $w = i_1 i_2 \ldots i_l \in \{1,2\}^*$, let

$$\bar{w} = (2-i_l)(2-i_{l-1}) \ldots (2-i_1).$$

Lemma 6. Let $w_1, w_2 \in \{0,1,2\}^*$, $w \in \{1,2\}^*$ and $\text{SEQ} = w_1 \cdot w \cdot 0 \cdot \bar{w} \cdot 0 \cdot w_2$.
Assume that Program 6 works on a plane R-ficograph of degree
bound 3 and with at least one $O3$-vertex. Let $l' = \text{len}(w_1)$
and $l'' = \text{len}(w_1 \cdot w \cdot 0 \cdot \bar{w} \cdot 0) = l' + 2 + 2 \cdot \text{len}(w)$.
Then the current position of the $(l''+1)$ st $O3$-macrostep is
equal to the current position of the $(l'+1)$ st $O3$-macrostep.

This follows from Lemma 5 by considering the behaviour of the
program at $O3$-vertices met during the working controlled by w.
Fig. 7 sketches the basic idea for $f = 1$ during the working
according to $w = i_1 i_2 \ldots i_l$.

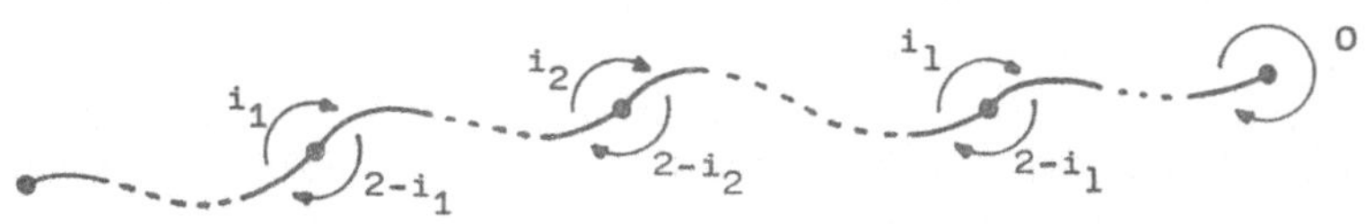

Fig. 7   (O3 – vertices)

Remark that if the control letter is equal to 0 at some O3-vertex,the value of f is changed. The value of variable s can remain unchanged. Indeed,when the program has a current vertex of order 2 at the first time after leaving an O3-vertex,its current position is of order 2 , i.e.,s obtains a new value,and the old one does not influence the behaviour. //

By Lemma 6,we can manage an automaton always to return to a certain basic position after arbitrary walks according to given control sequences w . Now we consider the problem to construct control sequences which enable the automaton to reach any vertex from a given position.

<u>Lemma 7.</u> Let h be a half-edge of order 1 in a plane R-ficograph L of degree bound 3. If L_h contains at most k O3-vertices , there is a word $w \in \{1\}^{*}$ with $\mathrm{len}(w) \leqslant k$ such that the following holds.
Assume that Program 6 working on L with $SEQ = w_1 \cdot w \cdot w_2$, for some $w_1,w_2 \in \{0,1,2\}^{*}$, has the current position h in the l th macrostep which begins with $z = \mathrm{len}(w_1) + 1$, $s = s' \in \{\text{left},\text{right}\}$ and $f = 1$. Then there is a number $l' > l$ such that $\mathrm{hal}(h)$ is the current position of the l' th macrostep and $z = \mathrm{len}(w_1 \cdot w) + 1$, $f = 1$, $s = s'$ at the beginning of this macrostep.

This can be proved by induction on the number of bridges in L_h.
If $\{h , \mathrm{hal}(h)\}$ is the only bridge in L_h,all vertices $\neq \mathrm{ver} \circ \mathrm{hal}(h)$ in L_h have the degree 2 or the order 3. Then in the macrostep following the l th one the program follows the face of ang(h) if its behaviour at O3-vertices is controlled by 1s only. Therefore,it will reach the current position hal(h) some time. Any O3-vertex of L_h can be only once the current vertex during this period of macrosteps,hence at most k 1s are needed in the control sequence. If $H(\mathrm{ver}(h)) = \{h , h_1 , h_2\}$ and $\mathrm{hal}(h_1)$ is the current position in the (l+1)st macrostep,then the program returns to ver(h) through h_2 , and we have $s = \mathrm{sid}_{h_2}(h) = s'$ in the next macrostep.

For the induction step,let $h' \notin \{h , \mathrm{hal}(h)\}$ be a half-edge of order 1 in L_h. If neither h' nor hal(h') is the current position during the walking through L_h controlled by 1s in O3-vertices,the behaviour of the program during this period is not essentially changed by removing the edge $\{h', \mathrm{hal}(h')\}$ together with that shore which does not contain h. From the validity of the assertion for the remaining labyrinth,the validity for L_h can be obtained.

Assuming,for example,that h' is the current position of some
of the considered macrosteps and L_h, does not contain the half-
edge h, the assertion follows from its validity both for L_h, and
for the labyrinth obtained from L_h by removing L_h. . //

<u>Lemma 8.</u> Let h be a half-edge with ord(ver(h)) = 3 and v be an
     arbitrary vertex of a plane R-ficograph L of degree bound **3**
     and with at most k O3-vertices. There is a sequence
     $w \in \{1,2\}^{\times}$ with len(w) $\leqslant$ k and satisfying the following
     property.
     If Program 6 with $SEQ = w_1 \cdot w \cdot w_2$, for some $w_1, w_2 \in \{0,1,2\}^{\times}$,
     has the current position h in the l' th macrostep,where
     $z = len(w_1) + 1$ and f = 1 at the beginning , then there is an
     l' th macrostep,for some l' $\geqslant$ 1,in which v is the current
     vertex and $z = len(w_1 \cdot w) + 1$.

     To défine such a sequence w , let
$$p = (h_0 , v_0 , h_0' , h_1 , v_1 , h_1' , \dots , h_m , v_m)$$
be a path connecting h with v in L , i.e. $m \in \mathbb{N}$, $h_0 = h$ and $v_m = v$.
We only consider the non-trivial case that m $\geqslant$ 1 and suppose
$v_i \neq v_j$ for 0 $\leqslant$ i < j $\leqslant$ m. Moreover,let p satisfy the

<u>Condition.</u> $h_0' \neq h_0$, and if ord(v_i) = 2 , ord(h_i) = 1 and j is the
     minimal number < i   such that ord(v_j) = 2 and ord(h_j) = 1,
     then $sid_{h_j}(h_j') \neq sid_{h_i}(h_i)$.

     The first part of the condition,$h_0' \neq h_0$,can be satisfied by
replacing a certain first segment ($h_0, v_0, \dots, h_i$) of an originally given path (violating this condition) by a path p' following
the border of the region corresponding to the face of ang(h_0),
see Fig. 8 .

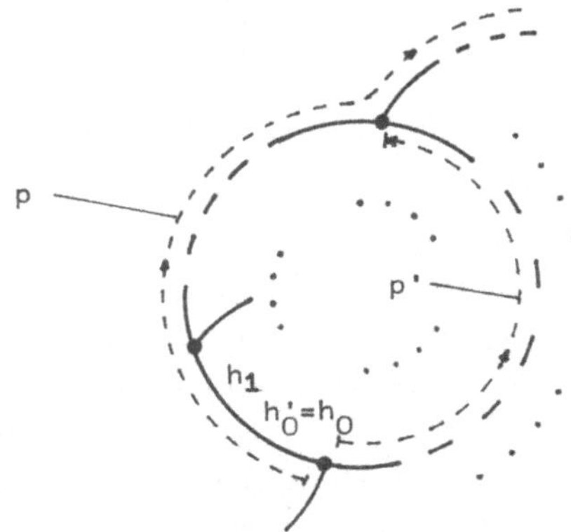

Fig. 8

By similar replacements,the second part of the condition can
be secured , see Fig. 9. Here,for all pairs (j,i) violating the
condition in p ,we replace certain segments (h_i' , ... , h_1) of p
by paths p'.

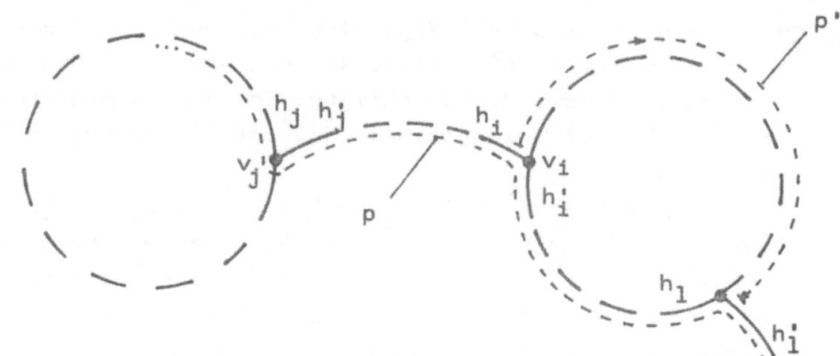

Fig. 9

Given a path p satisfying the condition,for $0 \leqslant i < m$ we define
control sequences $w(i) \in \{1,2\}^{*}$ leading from position h_i to
position h_{i+1} through the half-edge h_i'.

If $ord(v_i) = 3$,there is a $j \in \{1,2\}$ such that $h_i' = r^j (h_i)$,
and we define $w(i) = j$.

Now let $ord(v_i) = 1$. If $h_i' = r(h_i)$,we define $w(i) = \Lambda$.
Otherwise,we have $h_i' = r^2 (h_i)$. Then let $w(i)$ be the shortest
word in $\{1\}^{*}$ leading from $hal(r(h_i))$ to $r(h_i)$ according to
Lemma 7. $w(i)$ leads from h_i to h_{i+1} ,too,see Fig. 10.

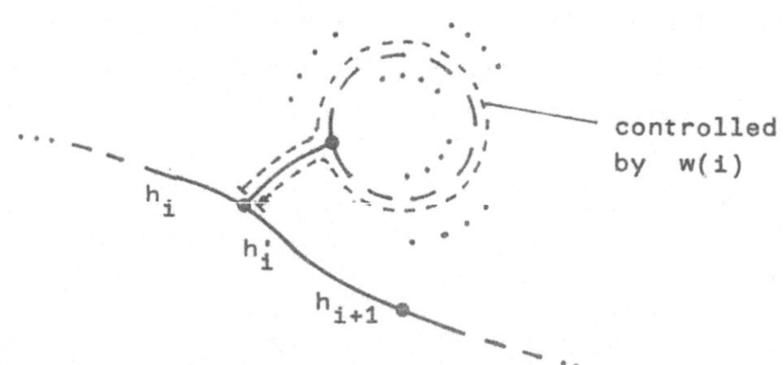

Fig. 10

Finally,we consider the case that $ord(v_i) = 2$. If $ord(h_i) = 2$
and $ord(h_i') = 1$,then let $w(i) = \Lambda$. For $ord(h_i) = ord(h_i') = 2$,we
define $w(i)$ like above as the shortest word in $\{1\}^{*}$ which leads
from $hal(h^{*})$ to h^{*} according to Lemma 7 , where $H(v_i) = \{h_i,h_i',h^{*}\}$.

If $ord(h_i) = 1$,then there is a maximal number $j < i$ such that
$ord(h_j) = 2$ and $ord(h_j') = 1$, since $ord(h_0) = 2$. We define $w(i) = \Lambda$,

102

and by the condition on the path p ,it follows that Program 6
goes from h_i to h_{i+1} in the corresponding macrostep.

It is easy to see that the sequence
$$w = w(0) \cdot w(1) \cdot \ldots \cdot w(m-1)$$
leads from the current position $h = h_0$ to h_m for $f = 1$, in the
sense of Lemma 8.

The words $w(i)$ are empty,or they control the behaviour of the
program in certain sublabyrinths of the form $L_{h_i^{\times}}$, where $hal(h_i^{\times})$
is incident to v_i and $ord(h_i^{\times}) = 1$. For $0 \le i < j < m$, $L_{h_i^{\times}}$ and
$L_{h_j^{\times}}$ have no common elements if they are defined. By Lemma 7 ,
$len(w(i))$ is bounded by the number of O3-vertices in $L_{h_i^{\times}}$, hence
$len(w) \le k$. //

Let $\left\{ w_i : 1 \le i \le 2^k \right\}$ be the set of all words of length k in
$\left\{ 1,2 \right\}^{\times}$, and
$$SEQ = w_1 \cdot 0 \cdot \overline{w_1} \cdot 0 \cdot w_2 \cdot 0 \cdot \overline{w_2} \cdot 0 \cdot \ldots \cdot 0 \cdot w_{2^k} \cdot 0 \cdot \overline{w}_{2^k} .$$
Then a 1-pointer automaton implementing Program 6 searches all
plane R-ficographs L of degree bound 3 with at most k O3-vertices.
For labyrinths L without O3-vertices,this already follows from
Theorem 4.

If there is an O3-vertex in L , then the automaton reaches such
a vertex $v^{\times}$ some time as the current vertex. During the following
walks corresponding to all possible control words of length k from
$\left\{ 1,2 \right\}^{\times}$, the automaton visits every vertex. This holds by Lemma 8,
whereas Lemma 6 secures that it always returns to the basic vertex
$v^{\times}$ and halts there finally.

Obviously,a 1-pointer automaton performing Program 6 with SEQ
as defined above needs $O(k \cdot 2^k)$ internal states.

<u>Theorem 6.</u> To any $b \geqslant 3$ and any $k \in \mathbb{N}$, there is a 1-pointer au-
tomaton which searches every plane R-ficograph of degree
bound b with at most k vertices of orders $\geqslant 3$.
This automaton has $O(b \cdot k \cdot 2^{b \cdot k}) = 2^{O(b \cdot k)}$ internal states,
and it eventually halts on labyrinths having at least one
vertex of an order $\geqslant 3$.

Let be given a plane R-ficograph L of degree bound b. The
plane R-ficograph $L^{\times}$ of degree bound 3 is obtained from L by
replacing the stars of vertices of degrees > 3 and orders $\neq 2$
by (mutually disjoint) hypervertices as sketched in Fig. 11a , and
by replacing the stars of vertices of degrees > 3 and order 2 by
hypervertices as sketched in Fig. 11b , where $ord(h_0) = ord(h_0') = 2$.
Stars of degrees ≤ 3 remain unchanged.

For an intuitive understanding of the construction of L^* the
reader is referred to Section 1.8 . Although L^* is not obtained
from L by a vertex substitution ('since the replacements depend on
the global property of order of vertices),the corresponding techniques can be immediately transferred.

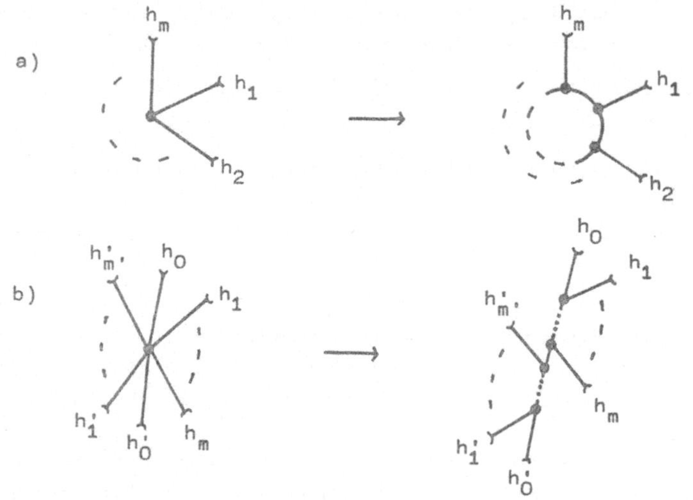

Fig. 11

L^* is of degree bound 3 and has at most b·k O3-vertices if L
has no more than k vertices of orders $\geqslant 3$. Let α^* be a 1-pointer
automaton implementing Program 6 and,therefore,searching all plane
R-ficographs of degree bound 3 and with at most b·k O3-vertices.

One easily constructs a 1-pointer automaton α which,on any
plane R-ficograph L of degree bound b,simulates the behaviour of
α^* on the labyrinth L^*. The number of states of α can be linearly bounded by the number of states of α^*. Therefore, α satisfies the assertion of Theorem 6. //

By the following lemma,the number of vertices of orders $\geqslant 3$ in
plane labyrinths can be linearly bounded by the number of faces.

<u>Lemma 9.</u> If a plane R-ficograph has exactly k vertices of orders
$\geqslant 3$ and m faces , then $k \leqslant 2m$.

To show this,we modify the given labyrinth by repeatedly removing vertices of degree 1 together with the incident edges and
removing vertices of degree 2 and joining the incident edges. By
this way,we finally obtain a plane R-ficograph having the same
numbers m of faces and k of vertices of orders $\geqslant 3$ as the original one. But every vertex has a degree $\geqslant 3$ now. Thus $3n \leqslant 21$

if n is the number of vertices and l the number of edges. By the
Euler theorem (Proposition 1.1), n - l + m = 2. It follows
3n ≤ 2n + 2m - 4 , hence k ≤ n ≤ 2m . //
   Using Lemma 9, from Theorem 6 we obtain

<u>Corollary 4.</u> To any b ⩾ 3 and any m ∈ $\mathbb{N}$, there is a 1-pointer au-
   tomaton which searches every plane R-ficograph of degree
   bound b with at most m faces. //

   Remark that there are labyrinths with arbitrarily many faces
but without vertices of orders ⩾ 3 , see Fig. 12.

Fig. 12

   As already mentioned in Section 1.6 , in C-ficographs the one
pointer can be replaced by a pebble. Taking into account the relationships between mazes and labyrinths as discussed in Section 1.4,
we obtain

<u>Corollary 5.</u> To any m ∈ $\mathbb{N}$, there is a 1-pebble automaton which
   searches every 2-dimensional finite maze whose order of connectivity is not greater than m . //

   The problem whether the one pointer in Theorem 6 can be replaced by one pebble is still unsolved. A replacement by two
pebbles is possible.

<u>Lemma 10.</u> To any degree bound b, there is a 2-pebble automaton
   which computes the order characteristic of its starting position h in plane R-ficographs in the sense of Lemma 2 if
   deg(ver(h)) ⩾ 3.

   The proof is rather simple. If necessary, the reader is referred
to /L.He86a/, pp. 254,255 . //

<u>Corollary 6.</u> The 1-pointer automata in Theorem 6 and Corollary 4
   can be replaced by 2-pebble automata. //

<u>Hints & Sources.</u> Theorem 5 has not been published so far. The
idea of results like Theorem 6 is due to K. Kriegel /L.Kr/,/L.HeKr/.
He proved Corollary 5 by methods applicable to 2D ficographs. The
generalization to plane R-ficographs and the concept of pointer
automata were given by the author /L.He84/,/L.HeKr/,/L.He86a/.
K. Kriegel (private communication) sketched a proof that the replacement of the pointer in Theorem 6 or Lemma 2 by a pebble is
possible for plane R-ficographs of degree bound three.

2.5. Edge-blocking in normed 2D ficographs

In this section we present a version of the edge-blocking
method for normed 2D ficographs. This yields a searching algorithm
which can be implemented by a counter automaton or a 2-pebble automaton or a system of two cooperating automata. The crucial point
is that the blocked edges are characterized by their coordinates,
and it is not necessary to mark them by special tokens.

Let be given a C-ficograph L and a normed 2D embedding P of L.

For any edge $e = \{h,h'\}$ of L, we define its coordinates (with
respect to P) by

$$coo(\,e\,) = \frac{1}{2} \cdot (\ P(\,ver(h)\,) + P(\,ver(h')\,)\,) \; .$$

This is the centre of the straight line segment connecting the
(embeddings of the) endpoints of e.

In the set of all edges of L we introduce an ordering $\preccurlyeq$ by

$$e_1 \preccurlyeq e_2 \quad \text{iff} \quad coo(\,e_1\,) \leqslant coo(\,e_2\,) \; ,$$

where the ordering $\leqslant$ in $\mathbf{R}^2$ is defined by

$$(\,x_1,y_1\,) \leqslant (x_2,y_2) \quad \text{iff} \quad \text{either} \quad x_1 < x_2$$
$$\text{or} \quad x_1 = x_2 \text{ and } y_1 \leqslant y_2 \; .$$

Obviously, $\preccurlyeq$ and $\leqslant$ are reflexive, transitive and antisymmetric,
and $\preccurlyeq$ does not depend on the choice of the normed 2D embedding
P of L.

We shall say that an edge $e = \{h,h'\}$ <u>belongs</u> to a face F of L
if ang(h') $\in$ F or ang(h) $\in$ F. Remark that e can belong to two different faces.

e is said to be the <u>minimal edge</u> of a face F if it belongs to F
and satisfies $e \preccurlyeq e'$ for all other edges e' belonging to F.
Obviously, for any face F we have exactly one minimal edge of F.

<u>Lemma 11.</u> Let e be the minimal edge of an interior face of a
normed 2D ficograph L. Then it is a vertical edge and belongs
to two different faces of L. If it belongs to two interior
faces, it is the minimal edge of only one of these.

If $e = \{h,h'\}$ would be a horizontal edge, the left of its endvertices would have degree 1, see Fig. 13a. Therefore, ang(h) and
ang(h') would be elements of the same interior face F of L. But
then the horizontal ray

$$\left\{ coo(e) + (x,-\tfrac{1}{2}\,) : \; x \leqslant 0 \right\}$$

must intersect the boundary of the interior region corresponding

to F in which it starts. The **first** intersection point would be
equal to coo(e') for some edge e' belonging to the same face F,
but with e' $\prec$ e .

Fig. 13

Now let e be a vertical edge as sketched in Fig. 13b. If the
point $coo(e) - (\frac{1}{2}, 0)$ would belong to the region corresponding
to the face F for which e is minimal,we would obtain a contradiction to the minimality. Therefore,if h is the upper half-edge
of e , ang(h) cannot belong to F. If the face of ang(h) is an
interior one,too, e cannot be the minimal edge of this face,as
follows by considering the ray
$$\left\{ coo(e) + (x, 0) : x \leqslant -\frac{1}{2} \right\} . \; //$$

<u>Lemma 12.</u> Let L be a normed 2D ficograph,and L' be obtained
   from L by removing all minimal edges of interior faces of L.
   Then L' is a normed 2D ficograph having exactly one face.

If L has no interior face,the assertion holds trivially. Otherwise,let L have m interior faces and $\{e_1, e_2, \ldots, e_m\}$ be the
set of all minimal edges of interior faces. By Lemma 11,all these
edges have the order 2.

Removing e_1 from L,we obtain a normed 2D ficograph having
exactly m-1 interior faces. Moreover, $\{e_2, \ldots, e_m\}$ is the set of
the minimal edges of these interior faces,as one easily shows.

Repeating this process of removing a minimal edge of an interior face,we finally obtain a normed 2D ficograph without interior faces , i.e. a normed 2D tree. //

<u>Lemma 13.</u> There is a 1-counter automaton which,in any normed
   2D ficograph L within $O(n_L)$ steps,decides whether its
   starting position belongs to the minimal edge of an interior
   and exterior face,respectively.
   More precisely,this automaton has internal states a_1, a_2, a_3, a_4,
   and,starting on a half-edge h of L,it halts after $O(n_L)$ steps
   in the state a, where
$$a = \begin{cases} a_1 & \text{if h does not belong to a minimal edge of a face ,} \\ a_2 & \text{if h belongs to the minimal edge of an interior} \\ & \text{face , but not of the exterior face ,} \end{cases}$$

$$
a = \begin{cases}
a_3 & \text{if } h \text{ belongs to the minimal edge of the exterior,} \\
& \text{but not of an interior face,} \\
a_4 & \text{if } h \text{ belongs to a minimal edge both of an exterior} \\
& \text{and an interior face.}
\end{cases}
$$

If $e = \{h_0, h_1\}$, this automaton checks for h_0 and h_1 whether e is the minimal edge of the face of $ang(h_0)$ and $ang(h_1)$, respectively, and, if this is the case, whether this face is an interior or exterior one.

To explain the basic idea, we assume that e is a vertical edge and h_1 is the lower half-edge, i.e., we consider the face to the right of e, see Fig. 14.

The automaton first follows the face of $ang(h_1)$ and uses the counter to store the difference between the x-coordinates of its current position and the edge e. More precisely, occupying a half-edge h in the face-following procedure, the counter keeps the value $z_h = x - x_0$ if $coo(e) = (x_0, y_0)$ and $coo(\{h, hal(h)\}) = (x, y)$.

Moreover, the internal state of the automaton is used to store the rotation index $rin(w_h)$ of the face-following path
$$w_h = (h_1 , ver(h_1) , r(h_1) , \dots , h) ,$$
which connects h_1 with h, modulo 3, i.e., it stores the remainder of $rin(w_h)$ divided by 3.

Let $z_{h^\times} = 0$ for some current position $h^\times$ and $z_h \geqslant 0$ for all positions h occupied in earlier steps of the face-following starting at h_1. Then $h^\times$ is the lower half-edge of a vertical edge $e^\times = \{h^\times , hal(h^\times)\}$. This lies below e iff $rin(w_{h^\times}) = 0 \equiv 0 \bmod. 3$, and $e^\times$ lies above e or is equal to e iff $rin(w_{h^\times}) = 4 \equiv 1 \bmod. 3$. This follows from Lemma 1.6, since the path $w_{h^\times}$ can easily be completed to a closed path giving the whole exterior or interior face, respectively, of a certain 2D ficograph, see Fig. 14.

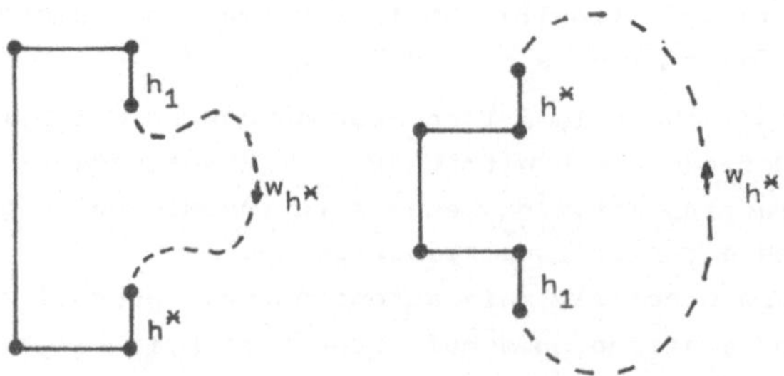

Fig. 14

If for the first occupied half-edge $h^\times$ that satisfies $z_{h^\times} = 0$ it holds that $rin(w_{h^\times}) \equiv 0 \bmod. 3$, then $e^\times \leqslant e$, i.e., e is not the minimal edge of the considered face. In this case,let the automaton return to the edge e with this information. e can be reached by going backward along the considered face until the x-distance z_h is equal to O again.

Now we assume that the half-edges h^* with $z_{h^*} = 0$ which are occupied in the course of the face-following perform a (possibly empty) sequence $h_1^*, \ldots, h_m^*$ such that $z_{h_i^*} = 0$, $rin(w_{h_i^*}) \equiv 1$ mod. 3 and $z_{h^*} > 0$ for all other positions h^* occupied in the face-following up to the reaching of h_m^*. Fig. 15 sketches this situation.

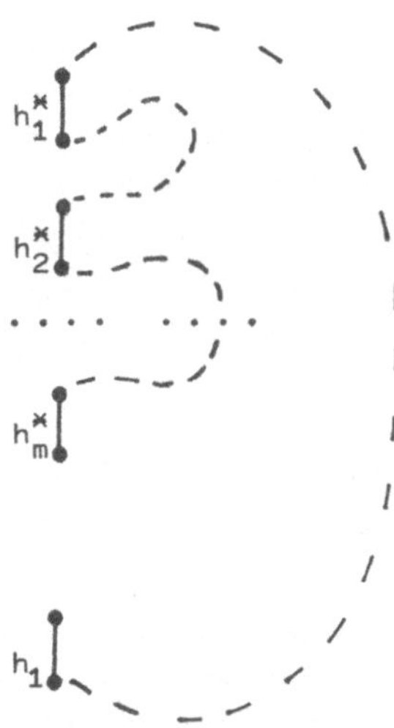

Fig. 15

If for the next position h^* that satisfies $z_{h^*} \leq 0$ it holds that $z_{h^*} < 0$, then $z_{h^*} = -\frac{1}{2}$, h^* belongs to a horizontal edge,and e is not the minimal edge of the face. Then the automaton returns to the edge e with this information. The returning is possible by following the considered face backward and counting z_h, and $rin(w_h)$ modulo 3. At position $h = h_1$ we have $z_h = 0$ and $rin(w_h) = 0$ mod 3 at the first time.

If none of the already discussed situations occurs,we finally obtain a sequence $h_1^*, \ldots, h_m^*$ as above but with $rin(w_{h_{m+1}^*}) \equiv 2$ mod 3 for the next position h_{m+1}^* satisfying $z_{h_{m+1}^*} = 0$ in the face-following. Then $h_m^* = h_1$ and $h_{m+1}^* = h_1^*$. In this case,the considered face is completely scanned,and e is its minimal edge. The automaton now returns to e with this information

simply by running backward along the face up to h_m^x ; that is the
first position h with $z_h = 0$.

So far we have described how to decide the property of minimality for a vertical edge e with respect to the face to the right-
hand side. If e is minimal with respect to this face,it must be an
interior one.

The case of considering the face to the left of a vertical edge
can similarly be treated,where in the first part of the procedure
the automaton may follow the face backward. If the edge will be
found to be the minimal,the corresponding face must be the exterior one.

Also the case of a horizontal edge can analogously be treated.
It is necessary for the minimality that the left endvertex of such
an edge has the degree 1,and in the case of minimality,the corresponding face must be the exterior one. //

<u>Lemma 14.</u> There are both a 2-pebble automaton and a system of two
        cooperating automata which always halt on their starting po-
        sition and decide whether it belongs to the minimal edge of
        an interior and exterior face,respectively,in a normed 2D
        ficograph L. By it,the 2-pebble automaton carries both pebbles
        and the two cooperating automata have the same position when
        they halt , and this is the case after $O(n_L^2)$ and $O(n_L)$
        steps,respectively.

To show this,we remark that the content z_h of the counter of
the automaton from the proof of Lemma 13 is bounded by card(F)
if F is the considered face. Therefore,this counter automaton can
be simulated by a 2-pebble automaton which represents the counter
content by the distance between the places of its pebbles along
the considered face. The first of the pebbles may represent the
position of the counter automaton,and the automaton moves like a
shuttle between its pebbles. Therefore,it needs $O(card(F)) = O(n_L)$
steps to simulate one step of the counter automaton.

Similarly,two cooperating automata can simulate the counter
automaton,where the counter content is represented by the distance
between the positions of the automata along the considered face.
The automata move with the basic velocity $\frac{1}{2}$ along the face. To

increase the counter,the first automaton (representing the position
of the simulated counter automaton) performs one step with the
velocity 1 ; to decrease the counter,it performs one step with the
velocity $\frac{1}{4}$. Thus one step of the counter automaton is simulated
by at most 4 steps of the cooperating system. //

Theorem 7. There are a counter automaton , a 2-pebble automaton
    and a system of two cooperating automata which search every
    normed 2D ficograph L and halt after $O(n_L^2)$, $O(n_L^3)$ and
    $O(n_L^2)$ steps,respectively.

This easily follows from Lemmas 12,13,14. To search the labyrinth L,the automata simulate the search by face-following of the
tree L' obtained from L by removing all minimal edges of interior
faces of L. Let the automata halt when they repeatedly scan the
minimal edge of the exterior face of L. //

Hints & Sources. The results and proofs of this section are completely due to M. Blum and D. Kozen /L.BlKo/.

2.6. Regular swinging in 2D ficographs

Here we introduce a searching algorithm which works in arbitrary 2D ficographs and can be implemented by a (1-pebble,1-counter)
automaton,by a 2-pebble automaton,or by a system of two cooperating
automata. The basic idea underlying this algorithm will be referred
to as the (regular) swinging method.

By a swing in a 2D graph L,we understand a face-following path
which starts at some vertex in the direction south , terminates at
some vertex with the direction north and has the rotation index 2.
More precisely,a path
$$w = (v_0 , h_0' , h_1 , v_1 , h_1' , \ldots , h_m , v_m)$$
with $v_0,v_1,\ldots,v_m \in V_L$ is a swing in L if $m \geqslant 3$, $h_i' = r_L(h_i)$ for
$1 \leqslant i < m$, $c_L(h_0') = c_L(h_m) = south$, and $rin(w) = 2$. Here V_L is the
vertex set, c_L the compass system and r_L the induced rotation
system of L.

In the sequel,paths w will simply be written in the form
$$w = (v_0 , v_1 , \ldots , v_m) ,$$
since the underlying graph is simple , cf. Section 1.3 .

For instance,the paths (a,b,c,d,e,f) , (c,d,e,f) , (k,j,i,h,g) ,
(g,h,i,j,k,l,m,c,b,a) are swings in the 2D ficograph shown in
Fig. 16.

The swing w is said to be regular if there is no proper in-
itial part $(v_0,v_1,\ldots,v_{m'})$,with $m' < m$, which already is a swing.

For instance , (m,c,a,b) , (c,d,e,f) , (a,b,c,d,e,f) in Fig. 16 are
regular swings, but (k,j,i,h,g) is not regular.

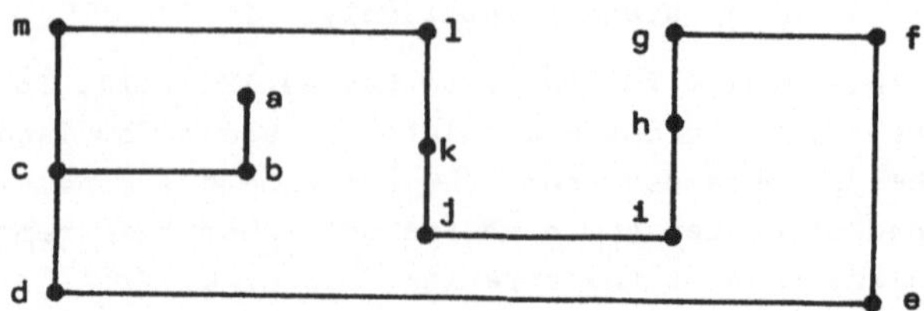

Fig. 16

Obviously, to every vertex v there is at most one regular swing
starting with v. By the following lemma, a regular swing cannot
traverse a whole face.

<u>Lemma 15.</u> If $(v_0, v_1, \ldots, v_m)$ is a regular swing in a 2D graph,
then $(v_i, v_{i+1}) \neq (v_j, v_{j+1})$ for $0 \leq i < j < m$.

Otherwise, we can consider the smallest integer i such that
there is an index j > i with $(v_i, v_{i+1}) = (v_j, v_{j+1})$. Since w is
face-following, we have i = 0 , and $(v_0, \ldots, v_j; v_{j+1} = v_0)$ gives a
face F of the considered labyrinth if j is minimal, too. By
Lemma 1.6 , $\mathrm{rin}(v_0, \ldots, v_j, v_{j+1}) = \overline{\mathrm{rin}}(F) \in \{4, -4\}$. Since
$w = (v_0, v_1, \ldots, v_m)$ is a regular swing, it follows that
$\mathrm{rin}(v_0, \ldots, v_j, v_{j+1}) = -4$. Moreover , $v_{j+k} = v_k$ for $0 \leq k \leq m-j$,
since w is face-following.

Therefore, we would have $2 = \mathrm{rin}(w) = \mathrm{rin}(v_0, \ldots, v_j, v_{j+1}) +$
$\mathrm{rin}(v_j, v_{j+1}, \ldots, v_m) = -4 + \mathrm{rin}(v_0, v_1, \ldots, v_{m-j})$. Thus,
$\mathrm{rin}(v_0, v_1, \ldots, v_{m-j}) = 6$; this contradicts the regularity of w. //

<u>Lemma 16.</u> Let , for some half-edge h of a 2D ficograph L,
$c_L(h) = $ south and ang(hal(h)) belong to an interior face.
Then there is a regular swing starting with vertex ver(h).

Indeed, let $v_0 = \mathrm{ver}(h)$, $h_0' = h$, $h_1 = \mathrm{hal}(h)$, and the path
$\overline{w} = (v_0, h_0', h_1, v_1, h_1', \ldots, h_{\overline{m}}; v_{\overline{m}} = v_0)$ give the interior face containing ang(h_1). By Lemma 1.6 , $\overline{\mathrm{rin}}(\overline{w}) = 4$, i.e. $\mathrm{rin}(\overline{w}) \geq 2$. Then
the smallest initial path w of $\overline{w}$ with rin(w) = 2 is a regular
swing starting with v_0 . //

A sequence $w' = (u_0, u_1, \ldots, u_{m'})$ is said to be <u>concatenable</u> with
the sequence $w = (v_0, v_1, \ldots, v_m)$ if $v_m = u_0$, and in this case, the

<u>concatenation</u> of w and w' is defined by
$$w \mathbin{\check{v}} w' = (v_0, v_1, \ldots, v_{m-1}, u_0, u_1, \ldots, u_m,) \ .$$
Obviously, this operation is associative.

If w and w' are paths, we have
$$rin(\,w \mathbin{\check{v}} w') = rin(\,w\,) + rin(\,w'\,) + rin(v_{m-1}, u_0, u_1) \ ,$$
hence the concatenation of two swings is a swing, too.

By a <u>wave</u> and a <u>regular wave</u> in a 2D graph, we mean the concatenation of finitely many swings and regular swings, respectively. For instance, $(k,j,i,h) \mathbin{\check{v}} (h,i,j,k,l,m,c,b,a) = (k,j,i,h,i,j,k,l,m, c,b,a)$ is a regular wave and $(a,b,c,d,e,f,g,h,i,j,k) \mathbin{\check{v}} (k,j,i,h) \mathbin{\check{v}} (h,i,j,k,l,m,c,b,a)$ is a non-regular wave in Fig. 16. Remark that the latter wave is closed.

<u>Proposition 1.</u> A regular wave w in a 2D ficograph cannot be
closed , i.e., if $w = (v_0, v_1, \ldots, v_m)$, then $v_0 \neq v_m$.

The proof is indirect. Assume that there is a closed regular wave in some 2D ficograph. Then let l be the smallest integer such that there is a closed regular wave being the concatenation of l regular swings in some 2D ficograph. Now let $w = w_1 \mathbin{\check{v}} \ldots \mathbin{\check{v}} w_l$ be a shortest closed regular wave of l regular swings $w_1, \ldots, w_l$ in some suitable 2D ficograph L .

<u>Claim.</u> Any swing w_i, $1 \leqslant i \leqslant l$, has the length 3 , i.e., it consists
of exactly three steps in the directions south, east and north, respectively.

This claim is indirectly proved, too. Without loss of generality, we assume that w_1 has a length greater than 3, say
$$w_1 = (v_0, v_1, \ldots, v_m) \ , \quad m > 3 \ .$$
Let the ficograph L' be obtained from L by inserting two new vertices $v_0^{\times}$, $v_m^{\times}$ into the edges $\{v_0, v_1\}$ and $\{v_{m-1}, v_m\}$, respectively, and connecting them by a horizontal edge, see Fig. 17 . (As usual in simple graphs, edges are determined by the sets of their endvertices.)

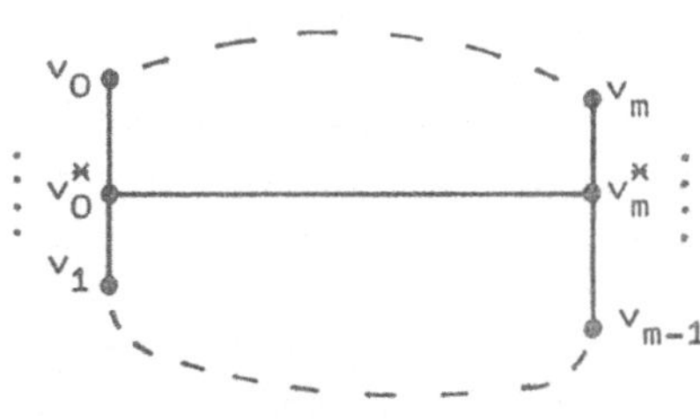

Fig. 17

Using Proposition 1.2 , one easily shows that L' is a 2D ficograph,since L is.

Now let w_i' be the uniquely determined regular swing in L'
with the same starting vertex as w_i, for $1 \leqslant i \leqslant l$.
So $w_1' = (v_0, v_0^{\varkappa}, v_m^{\varkappa}, v_m)$ has the length $3 < m$. For $1 < i \leqslant l$,if w_i
does not contain the directed edge (v_0, v_1) , then it does not con-
tain (v_{m-1}, v_m) , and $w_i' = w_i$. If (v_0, v_1) is contained in w_i,then
it also contains the whole swing w_1 , and w_i' is obtained from w_i
by replacing the part w_1 by w_1' . By Lemma 15,an edge can be contained at most once in a regular swing.

So we have obtained a closed regular wave
$$w' = w_1' \, \check{v} \, \ldots \, \check{v} \, w_l'$$
(in L') which is shorter than w . This contradiction proves the
claim.

Now the proof of Proposition 1 can easily be completed. From
the claim it follows that the wave w is of the form
$$w = (v_1', v_1, v_2, v_2', v_2, v_3, v_3', \ldots, v_1, v_1'),$$
see Fig. 18. Obviously,such a wave in a 2D ficograph cannot be
closed. //

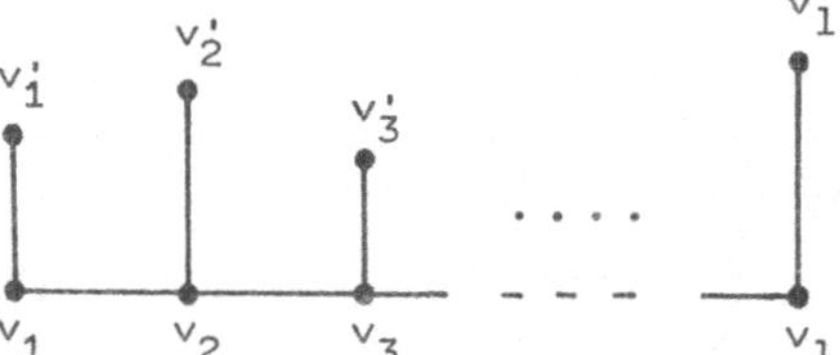

Fig. 18

An algorithm for reaching the exterior boundary in any 2D ficograph can already easily be described by means of our notations.
First,following the face of the starting position,one has to find
a southern directed edge. Then one has to continue the walk by
following the corresponding face until a regular swing is traversed. If (v_1, v_2) is the last directed edge of this swing,repeat
the regular swing-following stage starting with (v_2, v_1) , and so on
ad infinitum.

By Proposition 1 and since there are only finitely many regular
swings in a 2D ficograph,this walk finally becomes cyclic on some
face. By Lemma 16,this can be the case only on the exterior face.

To implement this algorithm,one needs a counter which keeps
track the (absolute value of the) rotation index of the traversed
path beginning with the first reached southern edge. During the
cyclic traverse of the exterior face,the content of the counter

exceeds any constant.

The only restriction of this method is due to the fact that
there are 2D ficographs having no vertical edge, see Fig. 19.

Fig. 19

Lemma 17. Let L be a 2D ficograph. There is a vertical edge in L
iff any face of L contains a southern edge.

More precisely, the condition means that any face contains an
angle ang(h), where $c_L(h) = $ north. One easily sees, if there is a
face in L without a southern directed edge, then all points from
$P(V_L)$ lie on some horizontal line in the plane, and conversely.
Here let P be a 2D embedding of L. //

Lemma 18. There is a finite automaton which, on any 2D ficograph L,
satisfies the following condition.
If there is no vertical edge in L, the automaton halts in a
special state a_- at the westernmost vertex after it has
searched the labyrinth ; otherwise it halts in a state $a_+ \neq a_-$
on some vertical edge.

This obviously follows from Lemma 17. //
The following program implements the idea described above.

Program 7.
object : a 2D ficograph L with the compass system c_L and the
$\qquad\qquad\qquad\qquad\qquad\qquad$ induced rotation system r_L ;

begin **if** there is no vertical edge in L
$\qquad\qquad$ **then** stop after searching L
$\qquad\qquad$ **else** move to some half-edge h with $c_L(h) = $ north ;
$\qquad$ z := 0 (: z keeps track the rotation index :) ;
$\qquad$ **loop** **let** h = current position ;
$\qquad\qquad$ **let** v_1 = ver(h) ; **let** v_2 = ver∘hal(h) ;
$\qquad\qquad$ **if** z = 2 **then** **let** $h^{\times}$ = hal(h)
$\qquad\qquad\qquad\qquad$ **else** **let** $h^{\times}$ = hal∘r(h) ;
$\qquad\qquad$ move to position $h^{\times}$;
$\qquad\qquad$ z := z + rin(v_2, v_1, ver($h^{\times}$))
$\qquad$ **repeat**
end

Theorem 8. A 1-counter automaton working according to Program 7
always reaches the exterior face in a 2D ficograph L. Here it
halts if L does not contain a vertical edge ; otherwise it
cyclically follows the exterior face without halting.

This is proved by the remarks made above and Lemma 18. //

Now we shall use the idea of regular swinging to obtain a searching algorithm.

Let ES_L denote the set of all southern directed edges of the given 2D ficograph $L = (V,H,I,c)$. These are pairs $(v,v') \in V^2$ such that there are half-edges $h,h' \in H$ satisfying the conditions $(v,h), (h,h'), (h',v') \in I$, $c(h) =$ south, and $c(h') =$ north. We shall simply say that the edge (v,v') _belongs_ to the face of $ang(h')$ Correspondingly, it can be an interior and an exterior edge, respectively.

For elements $e_1 = (v_1,v_1')$ and $e_2 = (v_2,v_2')$ from ES_L, we say that e_1 is a _son_ of e_2 or, equivalently, e_2 is a _father_ of e_1 if there is a regular swing

$$w = (u_0, u_1, \ldots, u_{m-1}, u_m)$$

such that $e_1 = (u_0,u_1)$ and $e_2 = (u_m,u_{m-1})$. This means, w starts with e_1 and terminates with the inverse edge of e_2.

The pair

$$F_L = (ES_L, \{ (e_2,e_1): e_1,e_2 \in ES_L \text{ and } e_1 \text{ is a son of } e_2 \})$$

is a (simple) directed graph in the usual sense. Fig. 20 shows a 2D ficograph L together with the corresponding graph F_L.

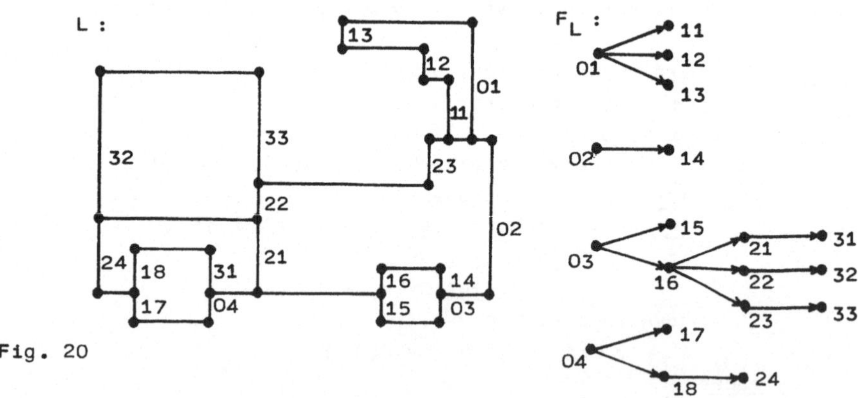

Fig. 20

It is not hard to show that F_L is a directed forest. Since, for any $e' \in ES_L$, there is at most one regular swing starting with e', the father of e' is uniquely determined if it exists.

Hence, for $e \in ES_L$, we obtain a sequence $e_0 = e, e_1, \ldots, e_k$ of edges $e_i \in ES_L$ such that e_{i+1} is a father of e_i ($0 \leq i < k$). If w_i is the regular swing starting with e_i , $w_0 \mathring{v} w_1 \mathring{v} \ldots \mathring{v} w_k$ is a regular wave in L. By Proposition 1, $w_i \neq w_j$, hence $e_i \neq e_j$ for $0 \leq i < j < k$. But there are only finitely many regular swings in L. Therefore,there is a maximal number k for which a sequence of the above described kind exists , and e_k has no father then.

We have shown that,for any $e \in ES_L$,there is exactly one directed path in F_L which terminates at e and starts at a uniquely deter- mined edge e^* having no father. This means that F_L is a directed forest.

Usually,the edges e^* which have no father are called the <u>roots</u> of F_L . By Lemma 16,all roots are exterior southern edges of the labyrinth L. But remark that not necessarily every exterior south- ern edge is a root.

<u>Lemma 19.</u> For any 2D ficograph L having a vertical edge, F_L is a directed forest whose roots are exterior southern edges. //

Now we introduce a relation between southern edges on the same level of F_L .

Let $e_1,e_2 \in ES_L$. Then e_1 is said to be an <u>elder</u> <u>brother</u> of e_2 or , equivalently , e_2 is a <u>younger</u> <u>brother</u> of e_1 if there is a face-following path $w = (v_0,v_1,\ldots,v_m)$ in L with $m \geq 2$, $e_1 = (v_0,v_1)$, $e_2 = (v_{m-1},v_m)$, $rin(w) = 0$ and $rin(v_0,v_1,\ldots,v_i) < 2$ for $0 < i < m$.

Obviously , $e_1 \neq e_2$ then , but e_1 and e_2 belong to the same face. If e_1 is a brother of e_2 and e_1 has a father,then e_2 has the same father.

<u>Lemma 20.</u> The relation " is a younger brother of " gives an irreflexive linear ordering in any class of southern edges having the same father and in the class of all southern edges having no father,respectively.
For $e_1,e_2 \in ES_L$, e_1 is an elder brother of e_2 iff there is a face-following path $w = (v_0,v_1,\ldots,v_m)$ with $m \geq 2$, $e_1 = (v_0,v_1)$, $e_2 = (v_{m-1},v_m)$, $rin(w) = 0$, $rin(v_0,v_1,\ldots,v_i) < 2$ for $0 < i < m$, and $(v_i,v_{i+1}) \neq (v_j,v_{j+1})$ for $0 \leq i < j < m$.

The proof of the second assertion is analogous to that of Lemma 15 ; it is omitted here.

Let e_1 and e_2 be southern edges with the same father,i.e.,the regular swings w_1 and w_2 starting with e_1 and e_2 ,respectively , terminate with the same edge. Then either $e_1 = e_2$ or e_1 is an elder or younger brother of e_2. By the second assertion of the lemma, e_1 cannot be both an elder and a younger brother of e_2. The transitivity of the relation is obvious.

Now let e_1 and e_2 be roots of F_L. Then they are exterior southern edges. Let $(v_0, v_1, \ldots, v_m ; v_{m+1} = v_0)$ give the exterior face, and $e_1 = (v_0, v_1)$, $e_2 = (v_i, v_{i+1})$ with $0 < i \leqslant m$. By Lemma 1.4 , $\mathrm{rin}(v_0, v_1, \ldots, v_i, v_{i+1}) \equiv 0 \ \mathrm{mod.}\ 4$, and $\mathrm{rin}(v_i, v_{i+1}, \ldots, v_m, v_0, v_1) \equiv 0 \ \mathrm{mod.}\ 4$.

We have $\mathrm{rin}(v_0, v_1, \ldots, v_i, v_{i+1}) \leqslant 1$, since e_1 would have a father otherwise. Analogously , $\mathrm{rin}(v_i, v_{i+1}, \ldots, v_m, v_0, v_1) \leqslant 1$. Since $\mathrm{rin}(v_0, v_1, \ldots, v_i, v_{i+1}) + \mathrm{rin}(v_i, v_{i+1}, \ldots, v_m, v_0, v_1) = \overline{\mathrm{rin}}(v_0, v_1, \ldots, v_m ; v_{m+1} = v_0) = -4$, it follows that
$$\left\{ \mathrm{rin}(v_0, v_1, \ldots, v_i, v_{i+1}) , \mathrm{rin}(v_i, v_{i+1}, \ldots, v_m, v_0, v_1) \right\} = \left\{ 0, -4 \right\}.$$
Therefore, e_1 is either a younger or an elder brother of e_2 . //

Our searching algorithm will follow the tree structure in the components of the forest F_L, where sons of the same father are treated in the succession from the elder to the younger ones. Also the treatments of the different components of F_L are ordered according to this succession between their roots. The algorithm will halt when the component with the youngest root has been repeatedly treated.

We consider the solutions of the following tasks by certain automata starting on an arbitrary southern directed edge e in a 2D ficograph L.

i) Decide whether e has an elder brother ; if yes, move to the youngest of all elder brothers of e , otherwise traverse the whole face of e and halt on e.

ii) Decide whether e has a younger brother ; if yes, move to the eldest of all younger brothers of e , otherwise halt on e after traversing the whole face of e.

iii) Decide whether e has a son ; if yes, move to the eldest son of e , but halt on e otherwise.

iv) Provided that e has no younger brother , decide whether e has a father ; if yes, move to the father , but halt on e otherwise.

<u>Lemma 21.</u> There are (1-pebble, 1-counter) automata performing the tasks i) - iv) within $O(n_L)$ steps. There are systems of two cooperating automata performing the tasks i) - iv) within

$O(n_L^2)$ steps. There are 2-pebble automata performing the
tasks i) - iv) within $O(n_L^3)$ steps. In all cases,these
decisions are indicated by the internal states of the automata , and when the automata halt,then they carry their pebbles
and both automata of a cooperating system have equal positions. The automata correctly work on any 2D ficograph L.

To decide whether its starting edge $e \in ES_L$ has an elder brother,
a (1-pebble,1-counter) automaton marks e by its pebble. Then it
backward follows the face of e keeping track in its counter the
rotation index of the face-following path w joining the current
position with e. When $rin(w) = 0$ and $rin(w') < 2$ for all proper
initial parts w' of w at the first time,then the southern edge $e^{\times}$
corresponding to the current position is the youngest elder
brother of e. Now the automaton moves back to e , picks up the
pebble and returns to $e^{\times}$.

If the edge e is reached after a complete walk through the face
of e without finding an edge $e^{\times}$ with the above property,then e
has no elder brother,as follows by Lemma 20. Then the automaton
halts on e with this information.

Task ii) can analogously be solved by following the face forward. To find the eldest son of e,the automaton backward follows
the face of the edge inverse to e ; Task iv) can be performed by
following the face of e forward. Obviously,all automata halt after
$O(n_L)$ steps.

To perform these tasks , 2-pebble automata or 2-automaton systems simulate a counter by representing its content by the distance between the pebbles and the automata,respectively,along the
corresponding face like in the proof of Lemma 14. Unfortunately,
then it is not possible to mark the starting edge e. But if there
is no elder brother,younger brother,son and father,respectively,
by Lemma 1.6,the content of the counter keeping track the (absolute
value of the) rotation index along the face will finally exceed
the number of edges in the face. This occurs after at most $O(n_L)$
complete walks around the face,i.e. after $O(n_L^2)$ steps of simulation , and can be realized by the situation that the pebbles or
the automata touch each other in such a way that the face segment
representing the counter becomes equal to the whole face. Then the
automata can return to the starting edge e by following their way
backward until the rotation index equals zero at the first time.
Remark that it is essential to suppose that e has no younger
brother,for Task iv) .

To complete the proof,remember that a 2-pebble automaton needs

$O(n_L)$ steps to simulate one step of the counter automaton as described above. //

<u>Program 8.</u>

```
object :  a 2D ficograph L ;
begin  if  there is no vertical edge in L
          then  stop after searching  L
          else  move to a southern edge in L ;
       z := 0 ;  d := 1 ;
       loop  let  h = current position ;
             let  e = ( ver∘hal(h) , ver(h) ) ;
             if  d = 1
                 then  if  e has a son
                           then  move to the eldest son of e
                           else  d := 0
                 else  (:  d = 0  now  :)
                       if  e has a younger brother
                           then  begin  move to the eldest
                                             ₀younger brother of e ;
                                       d := 1
                                 end
                       else  (:  e has no younger brother  :)
                             if  e has a father
                                 then  move to the father of e
                                 else  (: e is the youngest root :)
                                       begin  z := z + 1 ;
                                              if z = 2 then  stop ;
                                              d := 1 ;
                                              loop  if  e has no elder
                                                            brother
                                                        then  exit
                                                        else
                             begin  move to the youngest elder
                                           brother  of  e ;
                                    let  h = current position ;
                                    let  e = ( ver∘hal(h) , ver(h) )
                       end
                                              repeat
                                       end
       repeat
end
```

Let L contain a vertical edge. As already mentioned,our program
follows the tree structure of the components of F_L. The treatments of brothers are ordered by the elder-relation. The variable
d gives the direction of the search in the forest. $d = 1$ corresponds to the search for a son , $d = 0$ to the search for a younger
brother or for the father if there is no younger brother.

The variable z counts the multiplicity of the occurring of the
youngest root as the current position. The algorithm halts when
z becomes equal to 2. If the youngest root is the current position at the first time,then (according to the interior loop) the
program moves to the eldest root , and here a complete searching
of the labyrinth according to F_L begins.

Let the statements of Program 8 be performed according to the
proof of Lemma 21. Since any face of L contains a southern edge
which is the current position of some macrostep,all vertices are
entered by the automata,namely in the decision procedure for the
existence of a younger brother if the current position corresponds
to a youngest brother in this face.

Program 8 halts after $O(n_L)$ macrosteps. It performs only two
macrosteps with the youngest root of F_L as current position. The
number of steps within all the other macrosteps can be estimated
by the bounds given in Lemma 21. Multiplying these bounds by n_L,
we obtain bounds for the number of steps within macrosteps with
the youngest root as current position. Therefore,we have proved

<u>Theorem 9.</u>   There are a (1-pebble , 1-counter) automaton , a
2-pebble automaton and a system of two cooperating automata
which perform Program 8 and search any 2D ficograph L. They
halt after $O(n_L^2)$, $O(n_L^4)$ and $O(n_L^3)$ steps,respectively. //

The method of regular swinging leads to an interesting complexity measure for 2D graphoids L,namely the <u>rotation</u> <u>number</u>
defined by

$$\text{rot}(L) = \max \left\{ |\text{rin}(w)| \; : \; w = (v_0,v_1,\ldots,v_m) , m \geqslant 1 , \right.$$

is a face-following path with
$$(v_i,v_{i+1}) \neq (v_j,v_{j+1})$$
$$\left. \text{for } 0 \leqslant i < j < m \right\}.$$

<u>Theorem 10.</u>   To any $k \in \mathbb{N}$,there is a finite automaton which has
only $O(k)$ internal states , searches every 2D ficograph L
satisfying $\text{rot}(L) \leqslant k$, and halts after $O(n_L^2 \cdot k)$ steps.

To prove this,we have only to show the following lemma.

<u>Lemma 22.</u> To any k ∈ **N** ,there are finite automata which,in any
2D ficograph L satisfying rot(L) ≤ k,perform the tasks i) -
iv) ,halt after $O(n_L \cdot k)$ steps and have only $O(k)$ internal
states.

Like the 2-pebble automata or the 2-automaton systems in the
proof of Lemma 21 , let the finite automata simulate counter automata keeping track the rotation indices of their ways following
certain faces. When the absolute value of such a rotation index
would exceed k,the face-following can be broken off and the automata can return to their starting edge (with negative answers
to the decisions). By Lemma 1.6 , this will be the case after at
most $O(k)$ travellings through the whole face , i.e. after $O(n_L \cdot k)$
steps. //

<u>Hints & Sources.</u> The basic idea of regular swinging is due to
G. Asser /L.As/. His main result essentially corresponds to
Theorem 8. The general method was developed in /L.He86b/,where
Theorem 9 can be found. Remark,however,that there are slight
differences between the notations there and those used here.

2.7. Searching by means of space-bounded Turing tapes

There are only few results concerning upper bounds of the space
complexity of labyrinth problems , and these are rather simple.
For the definition of space-boundedness of Turing tape automata,
the reader is referred to Section 1.5 .

<u>Theorem 11.</u> For the class of all normed 2D ficographs,there is
a $\log(n_L)$ space-bounded Turing tape automaton searching any
labyrinth L of this type and halting after $O(n_L^2 \cdot \log(n_L))$
steps.

This follows by simulation of the counter automaton from the
proof of Theorem 7. Remember that the content of the counter of
this automaton is bounded by n_L. Therefore,representing this
content by its binary notation on the Turing tape,we obtain a
$\log(n_L)$ space-bounded Turing tape automaton. Unfortunately, to
simulate one step of the counter automaton,it may be necessary to
scan the whole inscription of the Turing tape which has the length
$O(\log(n_L))$. This leads to the time bound given above. //

To sketch a direct proof of Theorem 11 , we show

<u>Lemma 23.</u> There are $\log(n_L)$ space-bounded Turing tape automata
α_+ and α_- which, starting on a vertex v of a 2D ficograph L
with a normed 2D embedding P , satisfy the following conditions.

α_+ decides whether there is a vertex v' with $P_x(v') = P_x(v)$
and $P_y(v') > P_y(v)$, and it halts on the vertex v_+^* satisfying
$P_x(v_+^*) = P_x(v)$ and
$$P_y(v_+^*) = \min \left\{ P_y(v') : P_x(v') = P_x(v) \text{ and } P_y(v') > P_y(v) \right\}$$
if there is such one ; otherwise it halts on vertex v.
Here $P_x(v)$ and $P_y(v)$ denote the x- and y-coordinate of P(v),
respectively.
Analogously, α_- decides whether there is a vertex v' with
$P_x(v') = P_x(v)$ and $P_y(v') < P_y(v)$, and it halts on the vertex
v_-^* with $P_x(v_-^*) = P_x(v_-^*)$ and
$$P_y(v_-^*) = \max \left\{ P_y(v') : P_x(v') = P_x(v) \text{ and } P_y(v') < P_y(v) \right\}$$
in this case , but it returns to vertex v if v_-^* does not
exist.
Both automata halt after $O(n_L \cdot \log(n_L))$ steps .

If there is a northern neighbour v_0 of the starting vertex v
of α_+ , we have $v_+^* = v_0$.

Otherwise, let the automaton α_+ simulate a 3-counter automaton
which traverses the face corresponding to the region above P(v).
Two of the counters keep track the x- and y-distance between the
coordinates of P(v) and P(v') , where v' is the current position.
The third counter computes the minimal distance between the y-
coordinates of P(v) and all vertices v' belonging to this face
and satisfying $P_x(v') = P_x(v)$. After the face has been completely
traversed, the automaton can make the desired decision and enter
the vertex v_+^* if it is defined.

There are $O(n_L)$ steps of the 3-counter automaton in this
procedure , they can be simulated by $O(n_L \cdot \log(n_L))$ steps of the
Turing tape automaton.

The automaton α_- can analogously work. //

Using Lemma 23, we can construct a $\log(n_L)$ space-bounded automaton which , starting on some vertex v , enters all vertices
of the given labyrinth which have the same x-coordinate as v .
Finally, it is not hard to prove Theorem 11 in this way. (//)

<u>Theorem 12.</u> For the class of all normed 3D ficographs , there is
an (n_L) space-bounded Turing tape automaton searching all
labyrinths L of this type and halting after $O(n_L^3)$ steps.

To prove this, let a Turing tape automaton simulate the behaviour of a pointer automaton implementing Tarry's algorithm given by Program 1 (with respect to some rotation system induced by the compass system). Instead to mark the already visited vertices by pointers on the entering half-edges, this automaton may note a trace of its way through the labyrinth on its Turing tape.

More precisely, when k steps of the pointer automaton have been simulated and v_0 , v_1 , ... , v_k are the corresponding current vertices, let the Turing tape carry the word

$$w_k = (P(v_1) - P(v_0))(P(v_2) - P(v_1)) . . . . (P(v_k) - P(v_{k-1}))$$
$$\epsilon \left\{ (x,y,z) \epsilon \{0,1,-1\}^3 : \text{there is at most one component different from } 0 \right\}^x .$$

Here P denotes a normed 3D embedding of the given labyrinth. Using this trace, the simulation of the next step of Tarry's algorithm is possible. To do this by means of w_k, the automaton realizes which of the neighbours of v_k has been already visited, and it determines the entering edge for v_k. Then the corresponding Tarry action of entering v_{k+1} and constructing w_{k+1} can be performed.

Since the pointer automaton implementing Program 1 halts after $O(n_L)$ steps, we have $\text{len}(w_k) = O(n_L)$. The simulation of one step of the pointer automaton requires $(\text{len}(w_k))^2 = O(n_L^2)$ steps of the Turing tape automaton. //

The essence of the method used in this proof could be characterized as searching by <u>map construction</u> . Indeed, the Turing tape inscription w_{k^x} obtained after the last, say k^x th, step of the automaton can be regarded as a coding or map of the given labyrinth L.

Remark that this method can be transferred to any higher dimension if we define the normed d-dimensional compass ficographs in the straightforward manner, for $d > 3$.

So far in this section we have considered Turing tape automata with only one worktape. The use of automata with two or more worktapes does not improve the space bounds, but it may lead to a certain improvement of the time complexities. In this way, in Theorem 12 we can obtain the time complexity $O(n_L^2)$.

<u>Hints & Sources.</u> The results of this section can be regarded
to be folklore,more or less. For instance,the row by row searching
of normed 2D ficographs by a Turing tape automaton according to a
proof of Theorem 11 via Lemma 23 and a corresponding map construction were described by A. N. Shah /L.Sh73/. As will be shown in
Section 3.1 , for the cubic plane R-ficographs,there is no searching Turing tape automaton which is space-bounded by some function.
For the 2D ficographs this question is still open.

2.8. Searching all infinite connected 2D graphs

The denumerably infinite connected graphs are briefly referred
to as <u>incographs</u>. We shall show that there are systems of finitely
many cooperating automata able to search all 2D incographs. The
basic method for searching incographs is to <u>try</u> <u>all</u> <u>possible</u>
<u>finite</u> <u>ways</u> in any graph,always starting at and returning to some
basic position.

<u>Lemma 24.</u> There is a Turing tape automaton which searches all
    2D incographs.

Let $\{w_i : i \in \mathbb{N}^+\} = D_2^+$, where the sequence $w_1, w_2, \ldots$ corresponds to the lexicographic ordering in D_2^+, i.e. w_1 = east ,
w_2 = north , w_3 = south , w_4 = west , w_5 = east east , w_6 = east north ,
and so on.

There is a Turing tape automaton which successively generates
all words $\tilde{w_i} = w_i \overline{w_i}$ on its worktape,where $\overline{w_i} = \overline{d}_1 \overline{d}_{1-1} \ldots \overline{d}_1$ if
$w_i = d_1 d_2 \ldots d_1 \in D_2^+$. (Remember that $\overline{d}$ denotes the direction
opposite to d .)

Always after w_i has been generated,let the automaton under-
take a walk according to this control word in the given 2D incograph. That means,in the corresponding step,it moves in direction
d_j $(1 \leqslant j \leqslant 1)$ if this is possible in the labyrinth ; otherwise it
does not change its position but replaces $\overline{d}_j$ by § in $\overline{w_i}$. So it
always returns to its starting vertex finally.

Obviously,such an automaton searches any 2D incograph if it
starts from an arbitrary position. //

Remark that the given automaton searches all 2D ficographs,too.
By a slight modification,we obtain a Turing tape automaton which

searches all 2D incographs and ficographs and always halts on the
latter ones.

<u>Lemma 25.</u> There is a 2-counter automaton which searches all 2D
incographs and works in such a way that, always when it
changes its position in the labyrinth, one counter is empty.

This can be obtained from Lemma 24 using the standard technique
of the simulation of Turing machines by 2-counter machines, as
described by P.C.Fischer /S.Fi/ ; see also /S.Mi/. //

Now we shall simulate the 2-counter automaton from Lemma 25 by
a pebble automaton , a cooperating system and a multihead automaton,
respectively. Remember that 2D incographs may have only finite
faces, they may also have infinite faces embedded in bounded
regions of the plane, and they may have "unbounded" faces. Fig.21
shows all these types.

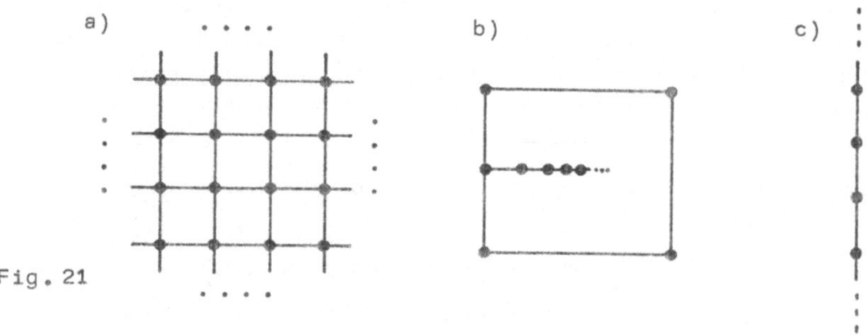

Fig. 21

<u>Theorem 13.</u> There are a 7-pebble automaton with distinguishable
pebbles, a system of 7 cooperating automata (even two cooperating automata with 5 distinguishable pebbles), and a 5-head
automaton which search every 2D incograph and every 2D ficograph and always halt on 2D ficographs after the search.

First we show that a 7-pebble automaton can search all 2D incographs. Because of the special kind of the 2-counter automaton in
Lemma 25, it is sufficient to show that three pebbles O , X_1 , Y_1
can mark the content of one counter, where pebble O lies on the
current position of the simulated counter automaton, and a pebble
automaton by means of two further pebbles X_w , Y_w is able to perform the counter operations (test for zero, increase or decrease
by one the counter content).

Then by means of two further pebbles X_2 , Y_2 and pebble O , the
second counter can be represented, and the automaton can simulate
a 2-counter automaton as long as its current position marked by

126

pebble O is not changed. But when this position must be changed,
one counter is empty,and by successively decreasing the one
counter value with respect to the position of O and simultaneously
increasing the other counter value with respect to the new position O' (which is a neighbour of the O-position) , the move of the
counter automaton to O' can be simulated. Finally pebble O is
moved to the new position.

To describe the implementation of one counter,we consider the
working of a 2-automaton system γ from Theorem 9,i.e. Program 8.
On any 2D incograph this system would infinitely long work without
a cycle in its behaviour. This can easily be shown. Therefore,the
number of steps performed up to a certain time (and the behaviour
of the system during these steps) is uniquely determined by the
starting position and the positions and states of the two automata
at this time. Finally,for any non-starting configuration,the previous configuration is uniquely determined. This is not trivial,
but follows from Program 8. We omit the details.

Now let O mark the starting position of the system γ, and X_1
and Y_1 the current positions of the two automata,whereas their
states are coded by the state of the pebble automaton. Then a
counter value is uniquely represented by the number of steps performed by γ from its starting position up to that configuration
marked by X_1 and Y_1 together with the corresponding states.
Moreover,by means of two further pebbles X_w,Y_w , the automaton can
find X_1 and Y_1 if it starts from O , and conversely ; and it can
increase and decrease the counter value and test it for zero.

We have shown how a 7-pebble automaton (with distiguishable
pebbles) can search any 2D incograph. If this automaton is put in
a 2D ficograph,it can realize this fact,since it reaches a halting
configuration of γ some time. This follows,because the values of
the counters of the automaton from Lemma 25 cannot be bounded by
a constant. When γ would halt,the whole labyrinth has been
searched , and the 7-pebble automaton may halt,too. This completes
the proof of Theorem 13 concerning the 7-pebble automaton.

One can show that the automaton together with the pebbles X_w
and Y_w in the above proof can be replaced by two cooperating automata each of which is able to see and to move the pebbles O ,
X_1 , Y_1 , X_2 , Y_2 . We omit the details of the proof here.

Finally,a multihead automaton can mark the positions of the
pebbles O,X_1,Y_1,X_2,Y_2 by five heads. Then it can directly simulate the counter operations , without using further heads or
pebbles. //

Remembering Theorem 10 , we obtain

<u>Theorem 14.</u>   To any $k \in \mathbb{N}$,there are a 3-pebble automaton with
distinguishable pebbles and a 3-head automaton which search
every 2D incograph L and 2D ficograph L satisfying
$rot(L) \leqslant k$ and always halt on such ficographs.

Indeed,in this case the configuration of a finite automaton
according to Theorem 10 can be marked by a single pebble X_1 or by
one head of a multihead automaton. Therefore,three pebbles O, X_1 ,
X_2 or three heads are sufficient to simulate the 2-counter automaton from Lemma 25 on the graphs of the given types. //

Now we shall show that also for normed 2D labyrinths the numbers of pebbles and cooperating automata in Theorem 13 can be
diminished. The corresponding searching algorithm is based on removing all minimal edges of finite faces of the given labyrinth.
The concept of minimal edge of a face has been introduced in
Section 2.5 for normed 2D ficographs,but it can immediately be
transferred to finite faces of normed 2D incographs.

<u>Lemma 26.</u>   Let L be a normed 2D incograph,and the normed 2D graph
L' be obtained from L by removing all minimal edges of finite
faces. Then there is no isolated vertex in L',and any face
of L' is infinite.

Any finite face F of L must be an interior one,i.e.,the corresponding region must be bounded,cf. Section 1.2 . To show this,
let P be a normed 2D embedding of L and
$$S_F = \bigcup \{ P(h) : h \text{ is a half-edge occurring in } F \}.$$
The complement of S_F in the plane consists of two regions. One of
these corresponds to the face F,the other one contains the embeddings of all vertices which do not occur in angles of F. Since
P is a normed 2D embedding of the 2D incograph L,the latter
region cannot be bounded,hence the former one is bounded.

Analogously to Lemma 11,we obtain that any minimal edge of an
interior face of L is vertical.

Let v be a vertex of L or L'. If there is a horizontal incident
edge in L,then this edge belongs to L',too. If there is no horizontal incident edge but there are two vertical incident edges in
L,then at least the upper one belongs to L',too. Finally,if there
is only one vertical incident edge in L,then it cannot be the
minimal edge of an interior face of L,as one easily sees. Therefore,v is not isolated in L'.

Now we assume that L' has a finite face F'. Let $F_1, \ldots, F_m$
denote all finite faces of L which touch vertices occurring in F',

and let w be a path in L which connects a vertex of F' with a
vertex $v^{\times}$ of L which does not occur in $F' \cup F_1 \cup \ldots \cup F_m$.

By L_0 we denote the sublabyrinth of L which is generated by the
vertices and half-edges occurring in w or $F' \cup F_1 \cup \ldots \cup F_m$. L_0 is
a normed 2D ficograph. If L_0' is obtained from L_0 by removing all
minimal edges of inner faces of L_0, then F' is a face of L_0' , as one
can show.

By Lemma 12 , L_0' is a 2D ficograph having exactly one face. But
the vertex $v^{\times}$ does not occur in angles of F'. This contradiction
shows that L' cannot possess a finite face. //

<u>Theorem 15.</u> There are a 5-pebble automaton with distinguishable
        pebbles and a system of five cooperating automata (even two
        cooperating automata with 3 distinguishable pebbles) which
        search every normed 2D incograph and ficograph and always
        halt on the latter ones.

To explain the basic idea of the algorithm,we first assume that
a normed 2D incograph L is given.

Remember that,in the proofs of Theorems 13 and 14,the values
of the counters (of an automaton from Lemma 25) have been represented by the numbers of work steps of certain automata starting
at a position marked by 0 up to configurations marked by further
pebbles. Now we represent the counter values by the numbers of
steps along certain infinite paths,so-called <u>counter paths</u>,in the
given labyrinth. The starting vertex is again marked by a pebble 0.
The crucial point is that also the end positions determining the
counter values can be marked by single pebbles X_1 and X_2,respect-
ively. The automaton together with two further pebbles,or two
cooperating automata,are used to manage the move along the counter
paths in order to simulate the counter operations.

On the class of all normed 2D incographs L which have only finite faces,the counter path w_v starting at a vertex v in L can
be defined as a certain face-following path in L' which starts
at v. By Lemma 26,such an infinite path w_v does exist for any
vertex v. The move along w_v in order to simulate the counter
operations can be performed by a finite automaton using two pebbles
or by two cooperating automata,as the proofs of Lemmas 13 and 14
show.

Unfortunately,in the general case L can possess infinite faces,
and there is no pebble automaton or cooperating system deciding
whether or not their starting position belongs to a finite face.
From this reason,the above idea is modified as follows.

Arriving at some vertex v_i,the counter path tries to traverse

all faces which are touched by v_i but have not yet been touched
in former steps of the counter path. The succession of these at
most four faces may be defined in some fixed way. If one of these
faces is infinite,we already have an infinite path to store any
possible value of the counter. If all these faces are finite,then
the counter path is continued by one step following the face in L'.

More precisely,the counter path starting at some vertex v can
be inductively defined. Let $v_0 = v$, h_0 be a half-edge incident to
v_0 and belonging to L'. h_0 can easily be determined,as shown in
the proof of Lemma 26. We define $w_0 = (v_0)$.

Assume that v_m , h_m and w_m are already defined for some $m \in \mathbb{N}$.
Then let $\hat{w}_m$ be the concatenation of face-following paths in L
which,in some fixed succession,traverses the L-faces in which the
vertex v_m occurs but no other v_i for $0 \leqslant i$ and $i < m$. If all these
faces are finite,then $\hat{w}_m$ starts and terminates at vertex v_m. If
such a face is infinite,then $\hat{w}_m$ becomes infinite,too.

In the latter case,if $\hat{w}_m = (v_m,v_{m+1},v_{m+2}, \dots)$, then let

$$w_v = (v_0 , v_1 , \dots , v_m , v_{m+1} , v_{m+2} , \dots) ,$$

i.e. the concatenation of w_m and $\hat{w}_m$,and the inductive process
of definition terminates.

In the former case,if $\hat{w}_m = (v_m,v_{m+1}, \dots ,v_{m+k} = v_m)$, we define
$h_{m+k+1} = hal \circ r_L.(h_m)$ and

$$w_{m+k+1} = (v_0 , v_1 , \dots , v_m , v_{m+1} , \dots , v_{m+k} , v_{m+k+1})$$

with

$$v_{m+k+1} = ver (h_{m+k+1}),$$

where $r_L.$ denotes the rotation system of the labyrinth L'. This
means , w_{m+k+1} is the concatenation of w_m , $\hat{w}_m$ and a face-following
step in L' from position h_m to position h_{m+k+1} .

Now a further induction step may be applied to v_{m+k+1} , h_{m+k+1}
and w_{m+k+1} .

If the inductive process never terminates by meeting an infinite face of L,then let w_v be that uniquely determined infinite
path having the paths w_i obtained by the induction steps as
initial segments. So the definition of w_v is complete.

As already mentioned,the current vertex v of the simulated
counter automaton is marked by a pebble O,and the value $z \in \mathbb{N}$
of the i th counter is represented by a pebble X_i on the vertex v_z
of the counter path $w_v = (v_0 , v_1 , \dots , v_z , \dots) .$

The vertex v_z can occur more than once in w_v ,namely as the
endvertex in some w_m, i.e.,in a step following the L'-face, or

130

within some $\hat{w}_m$, i.e., in some procedure traversing faces of L. By storing the information to which phase of w_v the value z of the counter corresponds, exactly one value z is determined by the pebble X_i on v_z. Moreover, if v_z belongs to some $\hat{w}_m$, the corresponding vertex v_m is uniquely determined. It is the first of the vertices visited in the basic L'-face-following routine and occurred in the L-face traversed in the phase corresponding to z.

It remains to explain how the operations of the counter can be simulated.

Starting at vertex v, an automaton using two further pebbles can traverse the initial segment of w_v up to the vertex v_z marked by X_i. To do this, the basic routine following the L'-face is interrupted at any visited vertex v_m by the procedure traversing all L-faces touching v_m, corresponding to the fixed succession of these faces. During this procedure, v_m is marked by a pebble. If one of these L-faces is infinite, the automaton finds the pebble X_i there. Otherwise, it is also possible that it finds the pebble X_i on such a face. If it has traversed all these faces without finding X_i, then they are finite, and the automaton using its two pebbles is able to simulate a further step of the basic routine following the L'-face.

In any case, after it has reached the position v_z, the automaton is able to decrease or increase by one the content of the counter by shifting the pebble X_i. If v_z is a position belonging to the basic routine and the content of the counter must be increased, it is necessary to decide which of the L-faces touching v_z have already been entered in former steps of w_v. To do this, the automaton follows the L'-face backward and traverses all L-faces touched in this routine. These are finite, and if the automaton meets the pebble X_i here, it stores the information that the corresponding L-face touching v_z has been searched. After reaching the starting position of w_v in this routine, the automaton returns to v_z again and performs the shift of X_i, corresponding to the increase of the value z by one.

After simulating a counter step, it is easy to find the vertex v by following the L'-face backward, possibly after the last previous position belonging to the basic routine has been entered.

So we have sketched the basic principle of the simulation of a 2-counter automaton by a 5-pebble automaton. Of course, to perform a step following the L'-face, the techniques described in the proofs of Lemmas 13 and 14 must be applied. This is possible, since the corresponding L-faces are recognized to be finite.

Instead of the finite automaton with two further pebbles,also
two cooperating automata can manage the counter simulation with
the pebbles O,X_1,X_2 on normed 2D incographs.

If the given labyrinth L is a normed 2D ficograph,our automata
can realize this fact,since the routine following the L'-face
becomes cyclic in this case,i.e.,the starting position marked by
pebble O will be reached in some step following the face of L'.
Then L has been completely searched,and the automata may halt. //

<u>Hints & Sources.</u> A searching algorithm for all normed 2D inco-
graphs was first given by M. Blum and W. Sakoda /L.BlSa/. It
could be implemented by a 7-pebble automaton with distinguishable
pebbles. This result was improved by A. Szepietowski /L.Sz82/,who
essentially proved Theorem 15. Using a result by J. Ulehla
(unpublished,cf. Section 4.2),the distinguishable (coloured)
pebbles in Theorem 15 can be replaced by indistinguishable
('uncoloured) ones.

A searching algorithm for all 2D incographs by a multipebble or
multihead automaton or a finite cooperating system,as given by
Theorem 13,has not yet been published so far.

CHAPTER III. TRAP CONSTRUCTIONS

Here we present unsolvability results of searching problems.
They are obtained by suitable trap constructions.

Let be given a normal labyrinth problem $(\mathcal{L},\mathcal{A},\mathcal{T})$, where the
task $\mathcal{T}$ means to search all labyrinths from $\mathcal{L}$, cf. Section 1.6 .
Then by a <u>trap</u> for an automaton $\alpha \in \mathcal{A}$, we understand a pair
$T = (L , h)$, where $L \in \mathcal{L}$ and h is a position in L such that α does
not search L if it starts with position h.

In some cases we estimate the size of the trap T , i.e. the
number of vertices of L, with respect to the number of internal
states of the automaton α. All traps will consist of simple
graphoids.

Sometimes,we shall also consider infinite labyrinths L which
are not mastered by a certain automaton starting on a position h.
Then $T = (L , h)$ is called a <u>mastering trap</u>.

3.1. Plane R-traps for finite automata and related types

In this section we only consider R-ficographs of degree bound three. We shall secure that the automata start on such positions that they reach only vertices of degree three during their working. Hence they obtain the constant input information 3 , i.e., they are working autonomously. For a finite automaton, then it follows that the output sequence controlling the move in the labyrinth becomes cyclic ultimately. This means, it has the form

$$f = w_0 \cdot w^\infty = w_0 \, www \ldots \in \{\S,0,1,2\}^{\mathbb{N}},$$

where $w_0 , w \in \{\S,0,1,2\}^*$.

Therefore, the problem of constructing traps for finite automata is reduced to the problem of constructing traps for such ultimately cyclic sequences, interpreted as control sequences for walking through R-graphs of degree bound three. This leads to the following definitions.

By a <u>control sequence</u>, we mean an infinite sequence $f \in \{\S,0,1,2\}^{\mathbb{N}}$. It is said to be <u>ultimately cyclic</u> if it is of the form $f = w_0 \cdot w^\infty$ with some $w_0, w \in \{\S,0,1,2\}^*$. The elements of $\{\S,0,1,2\}^*$ are called <u>control words</u> in this context.

Let L be an R-graphoid of degree bound three , h a half-edge in L and w a control word. The position entered in L after a walk starting with h and controlled by w is denoted by $pos(h,w)$. It is inductively defined by

$$pos (h , \Lambda) = h ,$$

and

$$pos (h , w \cdot x) = \begin{cases} pos(h,w) & \text{if } pos(h,w) \text{ is defined} \\ & \text{and } x = \S , \\ hal(\, r^x(\, pos(h,w) \,) \,) & \text{if } pos(h,w) \text{ is} \\ & \text{defined } , x \neq \S \text{ and} \\ & r^x(\, pos(h,x) \,) \text{ is a bound} \\ & \text{half-edge} , \\ \text{undefined} & \text{otherwise} , \end{cases}$$

where $w \in \{\S,0,1,2\}^*$, $x \in \{\S,0,1,2\}$, and r denotes the rotation system of L.

For a control word $w = x_1 \ldots x_m$, let $\overline{w} = \overline{x_m} \ldots \overline{x_1}$, where $\overline{\S} = \S , \overline{0} = 0 , \overline{1} = 2$ and $\overline{2} = 1$.

<u>Lemma 1.</u> Let L be an R-graphoid of degree bound three, h a half-edge of L and w, w_1, w_2 control words. Then $pos (h , w_1 \cdot w_2) = pos (pos(h , w_1) , w_2)$, and $pos (pos(h , w) , 0 \cdot \overline{w} \cdot 0) = h$ if all occurring expressions are defined.

This can be easily proved by induction. //

For a control word w , an R-ficograph L and a half-edge h in L, we shall say that the triple (w,L,h) satisfies Condition (C) if the following holds.

$$(\,C\,) \begin{cases} \text{pos}(h\,,w) = h \;;\; \deg(v) \in \{1,3\} \quad \text{for all vertices v of L,} \\ \text{but there is a vertex } v^{*} \text{ in L with } \deg(v^{*}) = 1 \;;\; \text{and} \\ \deg(\,\text{ver}\,(\,\text{pos}(h\,,w')\,)\,) = 3 \text{ for all initial segments w'} \\ \text{of w .} \end{cases}$$

The importance of Condition (C) for trap constructions, at least for control sequences of the form $f = w^{\infty}$, is obvious.

<u>Lemma 2.</u> For any $w \in \{\S,1,2\}^{*}$, there is a plane R-ficograph L with a half-edge h such that (w,L,h) satisfies Condition (C). One can secure that L has at most two faces. It has four vertices if $w \in \{\S\}^{*}$, but at most $2 \cdot \text{len}(w)$ vertices otherwise.

The case $w \in \{\S\}^{*}$ is trivial , then let L be the tree

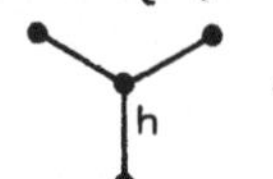

Without loss of generality, we now assume that $w = x_0 x_1 \ldots x_m$ with $m \in \mathbb{N}$ and $x_i \in \{1,2\}$ for $0 \le i \le m$.

Let the open plane R-graphoid
$$L' = (\,V'\,,H'\,,I'\,,r'\,)$$
be defined as follows , cf. Fig. 1.

$V' = \{\, v_i : 0 \le i \le m \,\}$;

$H' = \{\, h_i\,,h_i'\,,h_i'' : 0 \le i \le m \,\}$;

I' is the symmetric closure of the relation
$$\{\,(v_i\,,h_i)\,,(v_i\,,h_i')\,,(v_i\,,h_i'') : 0 \le i \le m \,\}$$
$$\cup \{\,(h_i'\,,h_{i+1}) : 0 \le i < m \,\} \cup \{\,(h_m'\,,h_0)\,\} \;;$$

and r' is a rotation system on $(\,V'\,,H'\,,I'\,)$ such that the restriction $r'/H(v_i)$ is the cyclic permutation $(h_i\,,h_i''\,,h_i')$ if $x_i = 2$, but $r'/H(v_i) = (h_i\,,h_i'\,,h_i'')$ if $x_i = 1$, for $0 \le i \le m$.

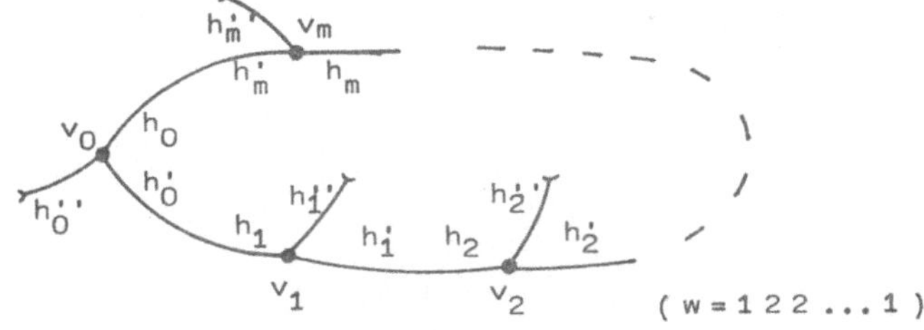

Fig. 1

$(\;w = 1\,2\,2\,\ldots\,1\;)$

It holds $\text{pos}(h_0, x_0 \dots x_i) = h_{i+1}$ for $0 \leqslant i < m$, and $\text{pos}(h_0, w) = h_0$.

Let L be obtained from L' by saturation of the free half-edges with (mutually disjoint) copies of the graphoid $\cdot\!\!\!\rightarrow\!\!\!\bullet$. Then for $h = h_0$ the required properties are satisfied. //

By a (plane) <u>3-trap</u> for a control sequence $f \in \{\S,0,1,2\}^{\mathbb{N}}$, we understand a pair (L, h) consisting of a (plane) R-ficograph L of degree bound three and a half-edge h of L such that

$$\deg(\text{ver}(\text{pos}(h,w))) = 3$$

for all initial segments w of f, and

$$\{\text{ver}(\text{pos}(h,w)) : w \text{ is an initial segment of } f\} \neq V_L.$$

Using Lemmas 1 and 2, we obtain plane 3-traps having at most two faces, for any ultimately cyclic 0-free control sequence f. If $f = w_0 \cdot w^{\infty}$, we construct the pair (L, h) according to Lemma 2 for the control word w. Replacing a suitable vertex of degree 1 in L by a sufficiently large tree, we obtain a plane R-ficograph Γ such that $\text{pos}_{\Gamma}(h, w')$ is defined for any initial segment w' of $0 \cdot \overline{w}_0 \cdot 0$ and is incident to a vertex of degree 3. For $h_0 = \text{pos}_{\Gamma}(h, 0 \cdot \overline{w}_0 \cdot 0)$, we have $\text{pos}_{\Gamma}(h_0, w_0 \cdot w^i) =$

$\text{pos}_{\Gamma}(\text{pos}_{\Gamma}(h, 0 \cdot \overline{w}_0 \cdot 0), w_0 \cdot w^i) =$

$\text{pos}_{\Gamma}(\text{pos}_{\Gamma}(h, 0 \cdot \overline{w}_0 \cdot 0 \cdot w_0), w^i) = \text{pos}(h, w^i) = h$, and all vertices visited during the walking according to $w_0 \cdot w^i$ have degree 3. Γ has at most two faces and contains at most $2 \cdot (\text{len}(w_0) + \text{len}(w))$ vertices if $w \notin \{\S\}^{*}$ or $\text{len}(w) \geqslant 2$.

The generalization of this construction to arbitrary ultimately cyclic control sequences intuitively seems to be rather simple. The control symbol 0 only causes to turn back some steps in the walking through labyrinths. Therefore, one should first construct a trap for a 0-free control sequence obtained from the originally given by a suitable reduction. Then this trap can be completed to a trap of the original sequence by replacing some vertices of degree 1 by sufficiently large trees.

To make the <u>reduction</u> of control sequences <u>modulo trees</u> precise, we introduce the following semi-Thue system of rules over the alphabet $\{\S,0,1,2\}$.

$$\mathcal{R} = \{ \quad \S \longrightarrow \Lambda ,$$
$$00 \longrightarrow \S ,$$
$$102 \longrightarrow 0 ,$$
$$201 \longrightarrow 0 ,$$
$$101 \longrightarrow 2 ,$$
$$202 \longrightarrow 1 \}.$$

For control words w and w', we write w ⊢— w' if there are
words w_0, w_0', w_1, w_2 such that $w = w_1 \cdot w_0 \cdot w_2$, $w' = w_1 \cdot w_0' \cdot w_2$, and
$w_0 \longrightarrow w_0'$ is a rule from $\mathcal{R}$. By $w \overset{k}{\vdash} w'$ for some $k \in \mathbb{N}^+$, we
mean that $w \vdash w_1$, $w_1 \vdash w_2$, . . . , $w_{k-1} \vdash w'$ for suitable
words $w_1, w_2, . . . , w_{k-1}$. $w \overset{0}{\vdash} w'$ is equivalent to w' = w .

<u>Lemma 3.</u> Let $w' \overset{k}{\vdash} w''$ for some control words w',w'' and $k \in \mathbb{N}$.
    Furthermore, let Condition (C) be satisfied by the triple
    (w'',L'',h''), where L'' is a plane R-ficograph.
    Then, by adding at most 2k vertices and edges to L'', it can
    be completed to a plane R-ficograph L' having as many faces
    as L'' such that Condition (C) is satisfied by (w',L',h'') .

    It suffices to prove the lemma for k = 1. Let $w' = w_1 \cdot w_0' \cdot w_2$,
$w'' = w_1 \cdot w_0'' \cdot w_2$ and $w_0' \longrightarrow w_0'' \in \mathcal{R}$, and (w'',L'',h'') satisfy
Condition (C).
    If $w_0' = \S$, then L' = L'' satisfies the assertion.
    For $w_0' = OO$, we consider the vertex $v = ver \circ pos_{L''}(h'', w_1 \cdot O)$
in L''. If $deg_{L''}(v) = 1$, we obtain L' from L'' by replacing the

graphoid $\bullet^v$ by the tree $\overset{v_1 \quad v_2}{\text{Y}}\,v$ with two new vertices

v_1, v_2 and corresponding edges. Then (w',L',h'') satisfies Condition (C), since $pos_{L'}(h'', w^*) = pos_{L''}(h'', w^*)$ for all initial
segments w^* of w_1, and $pos_{L'}(h'', w_1 \cdot w^*) = pos_{L''}(h'', w_1 \cdot O \cdot O \cdot w^*)$
for all initial segments w^* of w_2.
If $deg_{L''}(v) = 3$, we can take L' = L'' .
    For $w_0' = 102$ and $w_0'' = O$, let $v = ver \circ pos_{L''}(h'', w_1)$ and
$v' = ver \circ pos_{L''}(h'', w_1 1)$, see Fig. 2a . If $deg_{L''}(v') = 1$, we
replace v by a tree with two additional vertices v_1, v_2 as above ;
if $deg_{L''}(v') = 3$, let L' = L'' .
    The case $w_0' = 201$ can be analogously treated.

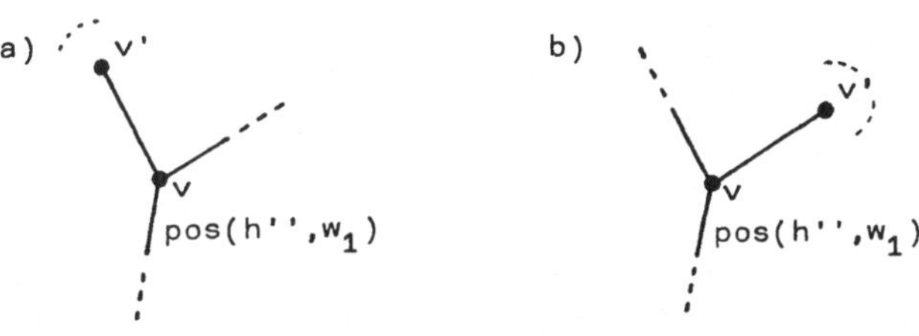

Fig. 2

Finally,we consider the case $w_0' = 202$; $w_0' = 101$ is analogous.
Let $v = \text{ver} \circ \text{pos}_{L''}(h'',w_1)$ and $v' = \text{ver} \circ \text{pos}_{L''}(h'',w_1 \cdot 2)$,
see Fig. 2b . We complete L'' to a labyrinth L' by adding two
edges and vertices connected with v' like above if $\deg_{L''}(v') = 1$;
if $\deg_{L''}(v') = 3$,then L' = L''. In any case,Condition (C) is satisfied by (w' , L' , h'') . //

Using Lemma 3,Lemma 2 could be generalized to arbitrary control
words w if we could show that any control word w can be reduced
to a 0-free control word $\hat{w}$,i.e., $w \xrightarrow{k} \hat{w}$ for some $k \in \mathbb{N}$. This
is not true,as the examples 0 , 01 , 10 , 020 show. But for our
purposes,the following weaker version is sufficient.

<u>Lemma 4.</u> Let $w \in \{\S,0,1,2\}^{\times}$. Then there is a number $k \leq 3 \cdot \text{len}(w)$
    such that one of the following properties holds.

    i) $w \xrightarrow{k} \hat{w}$ for some word $\hat{w} \in \{1,2\}^{\times}$;

    ii) $w^2 \xrightarrow{k} \Lambda$ or $w^3 \xrightarrow{k} \Lambda$;

    iii) $w = w_1 \cdot w_2$ for some words $w_1,w_2 \neq \Lambda$, and $w_2 \cdot w_1 \xrightarrow{k} \hat{w}$
         for some $\hat{w} \in \{1,2\}^{\times}$;

    iv) $w = w_1 \cdot w_2$ for some $w_1,w_2 \neq \Lambda$, and $(w_2 w_1)^2 \xrightarrow{k} \Lambda$
         or $(w_2 w_1)^3 \xrightarrow{k} \Lambda$.

To prove this,we first apply the rules of $\mathcal{R}$ to w (and the
words derived from w) as often as possible. We obtain $w \xrightarrow{k'} w'$
for some $k' \leq \text{len}(w)$,where w' satisfies one of the following
properties.

    (1) $w' = \Lambda$,

    (2) $w' \in \{1,2\}^{+}$,

    (3) $w' = 0$,

    (4) $w' = 0 \cdot w''$,

    (5) $w' = w'' \cdot 0$,

    (6) $w' = 0 \cdot w'' \cdot 0$,

with $w'' \in \{1,2\}^{+}$ in (4),(5) and (6).

For (1) and (2) , the assertion of Lemma 4 is satisfied. From
(3) it follows $w^2 \xrightarrow{2k'} 00 \xrightarrow{2} \Lambda$.

For Case (4) , we consider the subcases $\text{len}(w'') = 2l$,with
$l \in \mathbb{N}^{+}$, and $\text{len}(w'') = 2l+1$,with $l \in \mathbb{N}$.

4.1. Let $w'' = w_1'' \cdot w_2''$ with $\text{len}(w_1'') = \text{len}(w_2'') = l$. There are
control words w_1, w_2 such that $w = w_1 \cdot w_2$, $w_1 \xrightarrow{k_1} 0 w''$,
$w_2 \xrightarrow{k_2} w_2''$ and $k_1 + k_2 = k'$. Therefore, $w_2 \cdot w_1 \xrightarrow{k'} w_2'' \cdot 0 \cdot w_1''$.
Since $w_1'',w_2'' \in \{1,2\}^{+}$, we obtain either $w_2'' \cdot 0 \cdot w_1'' \xrightarrow{k''} \hat{w}$, for
some $\hat{w} \in \{1,2\}^{+}$ and $k'' < l$,or $w_2'' \cdot 0 \cdot w_1'' \xrightarrow{l} 0$. In the former

case,we have iii) from **Lemma 4**. The second case leads to
$(w_2{\cdot}w_1)^2 \overset{2k'}{\longmapsto} (w_2'{\cdot}0{\cdot}w_1')^2 \overset{21}{\longmapsto} 00 \overset{2}{\longmapsto} \Lambda$, i.e., we have the
property iv).

4.2. Let $w'' = w_1''{\cdot}x{\cdot}w_2''$ with $\mathrm{len}(w_1'') = \mathrm{len}(w_2'') = 1$ and $x \in \{1,2\}$.
There are control words w_1, w_2 such that $w = w_1{\cdot}w_2$, $w_1 \overset{k_1}{\longmapsto} 0{\cdot}w_1''$,
$w_2 \overset{k_2}{\longmapsto} x{\cdot}w_2$ and $k_1 + k_2 = k'$; hence $w_2{\cdot}w_1 \overset{k'}{\longmapsto} x{\cdot}w_2''{\cdot}0{\cdot}w_1''$.
If $x{\cdot}w_2''{\cdot}0{\cdot}w_1'' \overset{k''}{\longmapsto} \hat{w}$ for $\hat{w} \in \{1,2\}^+$ and $k''<1$,we have the
property iii). From $x{\cdot}w_2''{\cdot}0{\cdot}w_1'' \overset{1}{\longmapsto} x{\cdot}0$, it follows that
$(w_2{\cdot}w_1)^3 \overset{3(k'+1)}{\longmapsto} x0x0x0 \longmapsto \bar{x}0x0 \longmapsto 00 \overset{2}{\longmapsto} \Lambda$, i.e.,we
obtain the property iv).

The case (5) can be analogously treated.
Finally,we discuss the case (6). Here w can be represented as
$w = w_1{\cdot}w_2$ such that $w_1 \overset{k_1}{\longmapsto} 0$, $w_2 \overset{k_2}{\longmapsto} w''{\cdot}0$ and $k_1 + k_2 = k'$. It
follows $w_2{\cdot}w_1 \overset{k'}{\longmapsto} w''{\cdot}00 \overset{2}{\longmapsto} w''$,thus the property iii) is
satisfied. //

<u>Theorem 1.</u> For any ultimately cyclic control sequence $f = w_0{\cdot}w^\infty$,
   with $w_0, w \in \{\S,0,1,2\}^*$, there is a plane 3-trap (L,h) ,
   where L has at most two faces and $O(\mathrm{len}(w_0) + \mathrm{len}(w))$ vertices.

By Lemma 4,we obtain a representation $f = w_0'{\cdot}(w')^\infty$ such that
$\mathrm{len}(w_0') \leq \mathrm{len}(w_0) + \mathrm{len}(w)$, $\mathrm{len}(w') \leq 3\,\mathrm{len}(w)$ and $w' \overset{k}{\longmapsto} \hat{w}' \in \{1,2\}^*$,
where $k \leq 3\,\mathrm{len}(w)$.
By Lemma 2,there is a plane R-ficograph $\hat{L}$ with at most two
faces and $\leq \max\{4, 2\,\mathrm{len}(\hat{w}')\} \leq \max\{4, 6\,\mathrm{len}(w)\}$ vertices such
that $(w', \hat{L}, h')$ satisfies Condition(C)for a suitable half-edge
h' . Applying Lemma 3 , $\hat{L}$ can be completed to a plane R-ficograph
L' with at most two faces and $\leq 2\,\mathrm{len}(w') + 6\,\mathrm{len}(w) \leq 12\,\mathrm{len}(w)$
vertices such that Condition(C)is satisfied by the triple
(w',L',h') .
As already sketched for the 0-free case,by replacing some vertices in L' by sufficiently large trees,we obtain a plane R-ficograph L containing L' and having a half-edge h such that
$\mathrm{pos}_L(h,w_0') = h$, $\deg(\,\mathrm{ver}{\circ}\mathrm{pos}_L(\,h,w^*)\,) = 3$ for all initial segments w^* of $f = w_0'{\cdot}(w')^\infty$, but there are vertices of degree 1
in L. Moreover,L has not more faces than L' , and the number of
vertices increases by at most $2\,\mathrm{len}(w_0')$,i.e., L has at most
$2\,\mathrm{len}(w_0) + 14\,\mathrm{len}(w)$ vertices. //

139

Corollary 1. For any finite R-automaton with only m internal
     states , there is a trap (L , h), where L is a plane R-fico-
     graph of degree bound three,with at most two faces and O(m)
     vertices.

     Indeed,let $f = w_0 \cdot w^\infty$ be the ultimately cyclic control sequence
corresponding to the autonomous behaviour of the automaton if the
input signal constantly equals 3. One can secure that
$\text{len}(w_0)$, $\text{len}(w) \leqslant m$. Now let (L,h) be a plane 3-trap for f according to Theorem 1. //

Corollary 2. For any 1-counter automaton and even for any
     1-pushdown automaton there is a trap (L , h) , where L is a
     plane R-ficograph of degree bound three,with at most two
     faces.

     This follows from the fact that an autonomously working
1-pushdown automaton can only generate an ultimately cyclic output
sequence. For a detailed proof see /L.Co77,78/. //

     Remark that for the pushdown automaton in Corollary 2 the size
of the trap can be linearly bounded by the number of internal
states.

Corollary 3. Let $f: \mathbf{N}^+ \longrightarrow \mathbf{N}^+$. There is no f(n) space-bounded
     Turing tape automaton which searches all plane R-ficographs
     of degree bound three with at most two faces.

     We consider the behaviour of an f(n) space-bounded automaton
on the cubic R-ficograph with four vertices which is shown in
Fig. 3 a . Since the automaton can use only $k \cdot f(4)$ cells of its
worktape,for some constant $k \in \mathbf{N}^+$, it acts like a finite automaton.
The corresponding output sequence is ultimately cyclic. But this
is the output sequence corresponding to the autonomous behaviour
with the input symbol 3. Hence , Corollary 3 follows from
Theorem 1. //

a)

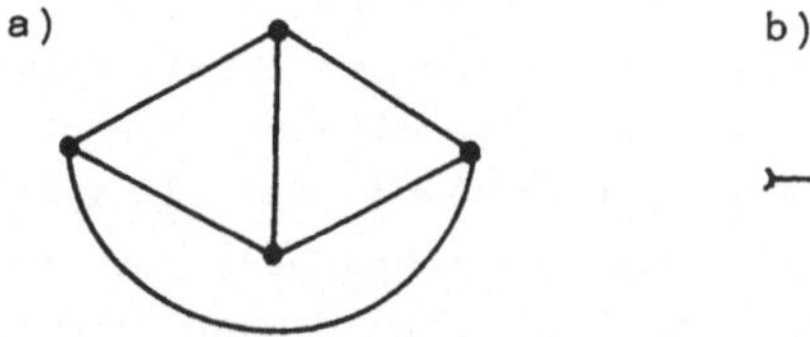

b)

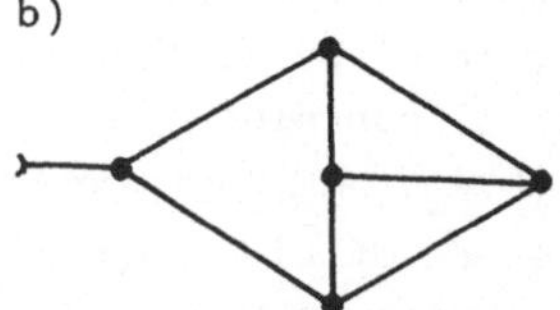

Fig. 3

     We remark that the stars of vertices of degree 1 in the traps
from Theorem 1 and Corollaries 1,2,3 could be replaced by (mutually disjoint) copies of the cubic plane R-ficographoid shown

in Fig. 3b . Then we obtain cubic plane traps. Of course,the
number of faces increases by this construction.

Now we shall sketch how to construct cubic plane R-traps whose
faces have bounded cardinalities. By Theorem 2.10 , such 2D traps
(for corresponding finite C-automata) cannot exist.

Theorem 2.   To any finite R-automaton,any 1-pushdown automaton
        and any f(n) space-bounded Turing tape automaton (where
        $f: \mathbb{N}^+ \longrightarrow \mathbb{N}^+$),there is a trap (L , h) such that L is a
        plane cubic R-ficograph in which every face consists of
        at most six angles .

From the proofs of Corollaries 2 and 3 it follows that it suffices to prove Theorem 2 for finite automata.

Let α be a finite R-automaton. It is started on the position
h_0 in the embedded cubic R-incograph L shown in Fig. 4a .
Remark that L is R-isomorphic to the embedded R-graph given by
Fig. 4b . We prefer L,since it enables us to use some straightforward notations.

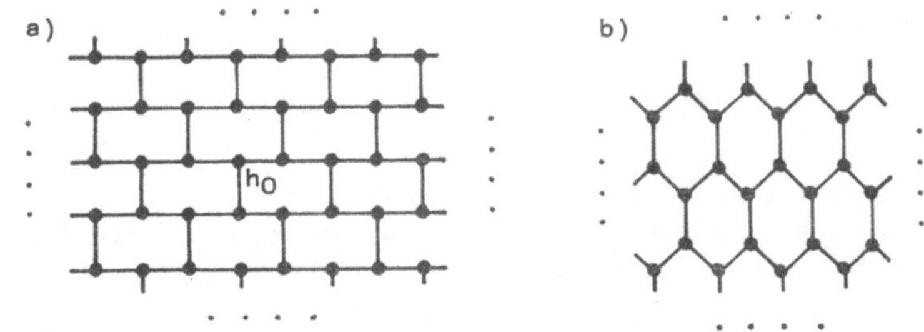

Fig. 4

We first assume that the automaton α starting on h_0 visits
only finitely many vertices of L . Then by cutting out a sufficiently large neighbourhood of h_0 in L , we obtain a trap. The
only problem is to cut suitable edges and to glue the arising
half-edges together in such a way that one obtains a plane cubic
R-ficograph with at most six angles in any face. Fig. 5 shows in
which way this can be done.

Now we assume that the automaton visits infinitely many vertices of L,but it crosses only finitely many horizontal strips.
Then there are two half-edges h_1 and h_2 entered by α and belonging to the same horizontal line in L such that the following
conditions hold.

(1) The horizontal translation of the plane from (the embedding

of) h_1 to (that of) h_2 causes an **R-isomorphism** of Γ onto
itself.

(2) Starting on h_0 , the automaton α has the same state both on
h_1 and on h_2 when it occupies these positions at the first
time.

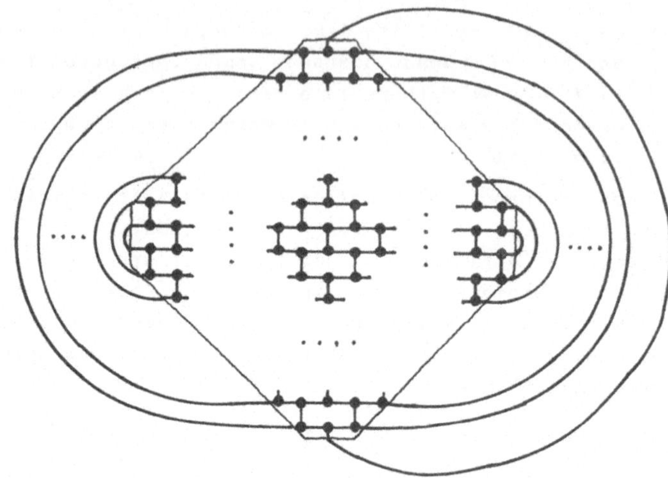

Fig. 5

Let δ_h be the distance between h_1 and h_2 , this is the minimal
number of vertices in paths connecting h_1 and h_2. $\delta^\times$ denotes the
number of horizontal lines of Γ which are crossed by α. If α
has only m states,then $\delta^\times \leq m$. If h_1 and h_2 are suitably chosen,
we have $\delta_h \leq 2\,m$. Moreover,$\text{pos}_\Gamma (h_1$,w$) = h_2$ if h_1 is first
entered by α and w is that segment of the output sequence which
leads α from h_1 to h_2. The whole output sequence has the form
$f = w_0 \cdot w^\infty$, for some word w_0 . Remark that len(w) $>$ m is possible
because of Condition (1).

We obtain the desired labyrinth L by cutting out a suitable
rectangular part of length $2\delta_h$ and height $\overline{\delta}^\times$ from Γ ,building
a cylinder then by connecting the free horizontal half-edges and
gluing the vertical free half-edges together,as shown in Fig. 6.
Here let $\overline{\delta}^\times$ denote the smallest even integer greater than $\delta^\times$.
Then we have $\text{pos}_L(h_1 , w^2) = h_1$. The starting position h in L
can be obtained as $h = \text{pos}_L(h_1 , 0 \cdot \overline{w}_0 \cdot 0)$.

Finally,we have to discuss the case that α crosses infinitely
many horizontal strips. Then there are two vertical half-edges h_1
and h_2 touching different horizontal lines in Γ such that

142

(1) the translation of the plane which maps h_1 onto h_2 causes an R-isomorphism of $\overline{L}$ onto itself, and

(2) α enters h_1 and h_2 in this succession and with the same state at the first time.

Let δ be the number of horizontal lines between h_1 and h_2. Fig. 7 shows the basic principle to construct a trap with the properties required in Theorem 2.

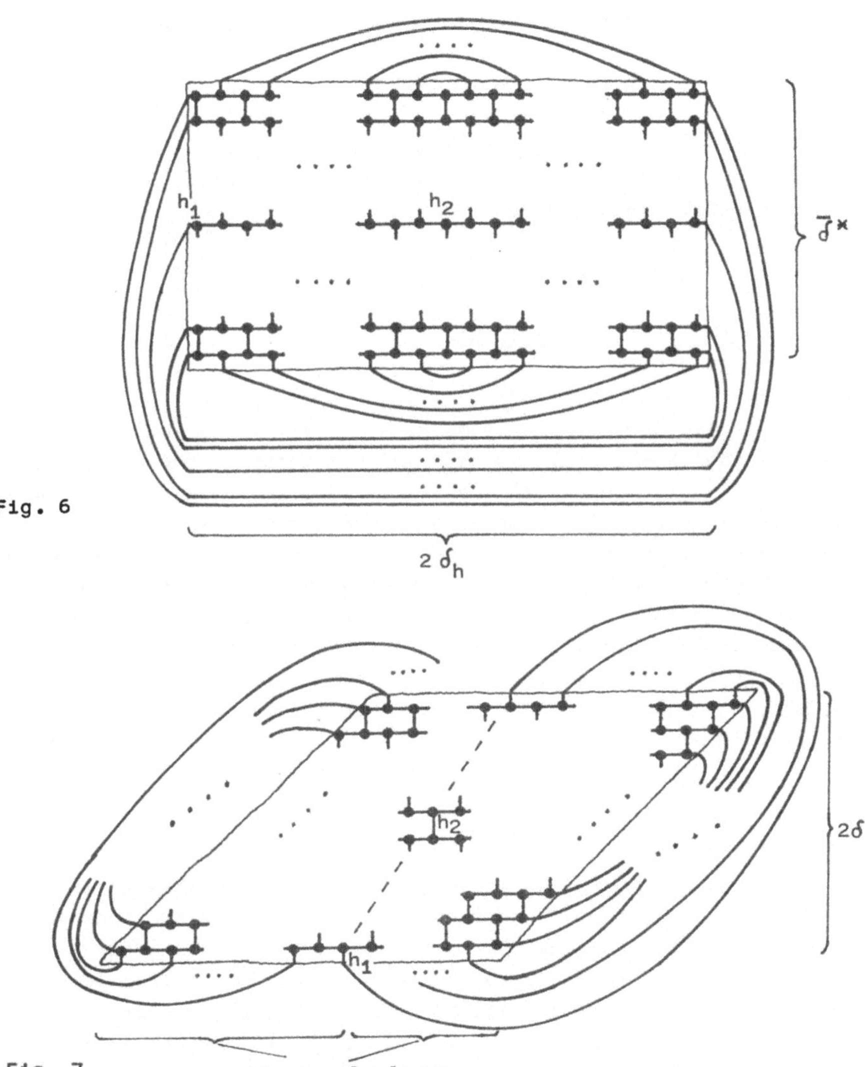

Fig. 6

Fig. 7

Remark that in all the cases the number of vertices of the
trap labyrinth L can be bounded by $O(m^2)$, where m is the number
of states of the finite automaton α. //

<u>Hints & Sources.</u> The basic ideas of the proofs of Theorems 1
and 2 are due to H. Müller /L.Mu71/. But he neither mentioned the
estimation of the size of the trap nor remarked the bound for the
number and the cardinality of faces,respectively. Our proof of
Theorem 1 follows the elaboration by M. Bull /L.Bul/,who especially found Lemma 4.
Corollary 2 is due to W. Coy /L.Co77,78/. Corollary 3 has been
remarked in /L.He82/. It seems to contradict Coy's result that
there is a linearly bounded automaton searching all R-ficographs.
His definition of linear boundedness,however,essentially differs
from the usual meaning of space-boundedness.

3.2. Traps for halting automata

In this section we shall deal with the searching abilities of
finite automata which always halt on labyrinths of the considered
type. The importance of halting searching algorithms is based on
the fact that they allow the decision of certain (locally determined) properties of finite labyrinths,whereas arbitrary searching
algorithms only yield recognition procedures,cf. Section 4.4 .
In the previous section we have constructed plane 2-face traps
for finite R-automata. It is a famous open problem whether also
for finite C-automata there exist 2D traps with only two faces,
cf. Section 4.1 . For halting finite automata,an affirmative
answer is well-known.

<u>Theorem 3.</u> There is no finite C-automaton which searches any
strongly normed 2D ficograph having at most two faces and
always halts at some time on these labyrinths.

By Lemmas 1.13 and 1.16 which can be easily generalized to
halting automata,it suffices to prove Theorem 3 for 2D ficographs.
Let a finite automaton starting on position h halt after k
steps on the rectangular 2D graph shown in Fig. 8 a . Then we consider the spiral shown in Fig. 8 b with at least k+1 edges in
both of the arms starting from ver(h). If it is started on the
position h in this spiral,the automaton also halts after k steps.

without searching the spiral. //

We remark that Theorem 3 and the proof given above can be immediately generalized to automata with auxiliary storage tapes.

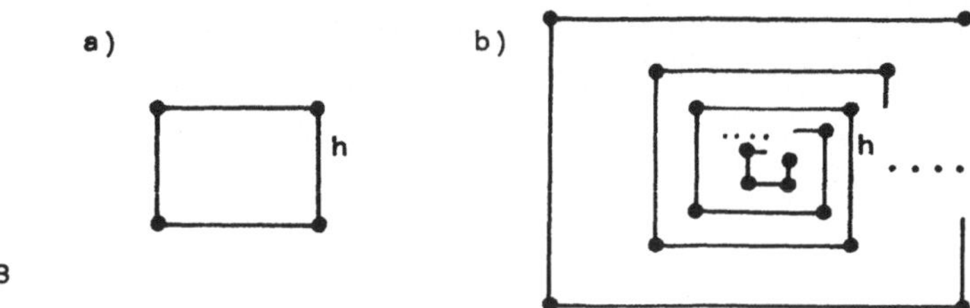

Fig. 8

In the remaining part of this section we shall prove the much stronger result that a halting finite automaton cannot search all finite trees of the usual types,even if they satisfy a certain regularity condition.

A graphoid or labyrinth is said to be <u>quasi-regular</u> (of degree b) if all its vertices v have degrees from $\{1,b\}$,i.e. $\deg(v) \in \{1,b\}$.

Remark that a quasi-regular connected graph of degree 1 is regular and consists of two adjacent vertices. The quasi-regular finite trees of degree 2 are chains with two or more vertices as sketched in Fig. 9 . Of course,labyrinths with such underlying graphs can be easily searched by halting finite automata.

Fig. 9

Theorem 4. Let b⩾3. There is no finite R-automaton which searches any quasi-regular finite R-tree of degree b and always halts at some time after the search.

Remark that any finite R-tree is plane.

The proof of the theorem is based on four lemmas. First we generalize Theorem 3.

Lemma 5. Let α be a finite R-automaton or an automaton with an auxiliary storage tape such as a Turing tape,a pushdown store or a stack. Suppose that α searches all (quasi-regular) finite R-trees of a certain degree bound. Then,starting with an arbitrary position in a (quasi-regular) R-ficograph which contains a cycle, α never halts.

The meaning of the brackets here is that the lemma holds both
for the quasi-regular and for the general case. We also remark
that it analogously holds for compass labyrinths.

Assume that L is a (quasi-regular) R-ficograph containing a
cycle and h_0 is a position in L such that α starting with position h_0 in L halts after k_0 steps.

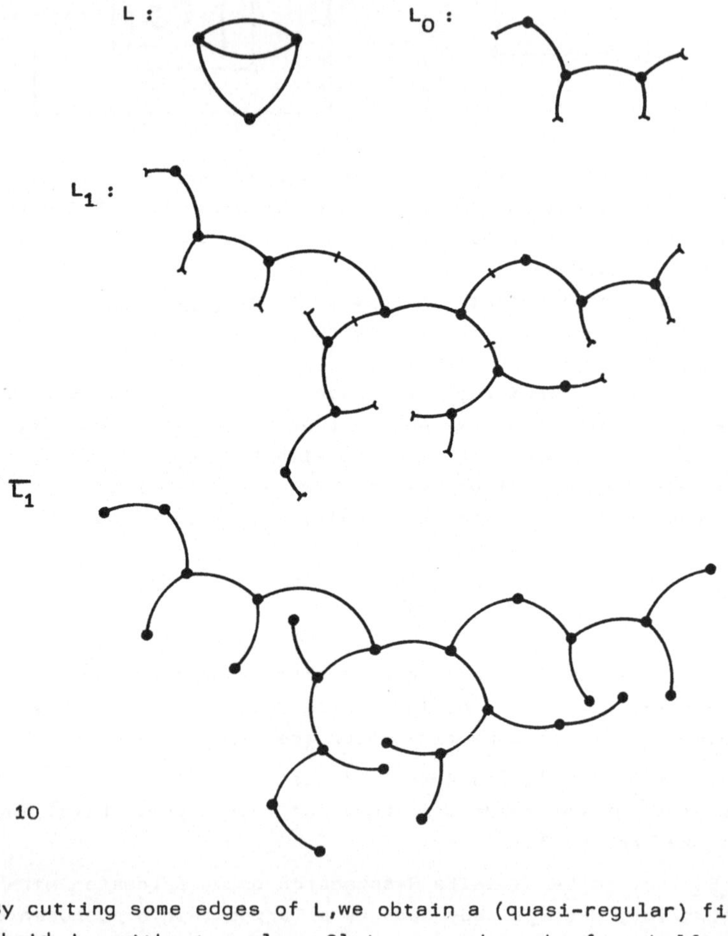

Fig. 10

By cutting some edges of L,we obtain a (quasi-regular) ficographoid L_0 without cycles. Gluing together the free half-edges
h of L_0 with the images of their counterparts hal(h) in pairwise
disjoint isomorphic copies of L_0,we obtain L_1. Generally,from L_k
we obtain L_{k+1} by gluing together the free half-edges of L_k
with the images of their counterparts in pairwise disjoint

146

isomorphic copies of L_0. **Fig. 10** illustrates this process.
Finally, $\overline{L_{k_0}}$ is obtained from L_{k_0} by saturation of the free
half-edges by pairwise disjoint copies of the R-graphoid ●—ㅜ .

Starting with position h_0 in $\overline{L_{k_0}}$, the automaton α receives
the same input information as in L , in any work step as long as
it does not reach an endvertex arised as a result of the satura-
tion process. Hence α halts after k_0 steps in $\overline{L_{k_0}}$, but then it
has not completely searched this finite R-tree. //

Let $\alpha = (X , Y , A , \delta , \lambda , a_0)$ be a finite R-automaton of
degree bound b , i.e. $X = \{1,2,\ldots,b\}$ and $Y = \{§,0,\ldots,b-1\}$.
α is said to be a <u>forward</u> <u>automaton</u> if
$$\lambda (x , a) \in \{1 , \ldots , x-1 \}$$
for any $x \in \{2,\ldots,b\}$ and any state $a \in A$. This means that the
automaton,arriving at a vertex of degree 1 , never goes back.

A state $a \in A$ is said to be <u>cyclic</u> if there is a number $k \in \mathbb{N}^+$
such that
$$\delta^*(\underbrace{b \ldots b}_{k \text{ times}} , a) = a .$$
(Here let δ^* denote the natural generalization of δ to the set
$X^* \times A$. It is inductively defined by $\delta^*(\Lambda , a) = a$ and
$\delta^*(w \cdot x , a) = \delta (x , \delta^*(w,a))$ for $w \in X^*$, $x \in X$, $a \in A$.)
If the state a is cyclic,then there is a number $k \leq \mathrm{card}(A)$
which satisfies the above equation.

<u>Lemma 6.</u> To any $m \in \mathbb{N}^+$ and $b \geqslant 3$, there is a regular R-corridor F
of degree bound b such that , for every forward automaton α
with at most m states and for every cyclic state a of α ,
$\varphi_{\alpha,F} (a , 1) = (a , 2)$ and $\varphi_{\alpha,F} (a , 2) = (a , 1)$.

To prove this,let T_k denote the complete open R-tree of
degree b and depth k ,where the free half-edges except the root
are labelled by the control words $w \in \{1,\ldots,b-1\}^k$ which lead
from the root to them ; see Fig. 11 for $b = 3$ and $k = 3$. Let $\overline{T_k}$
be a disjoint isomorphic copy of $\overline{T_k}$.

The desired corridor F can be obtained by gluing together the
free half-edge labelled with w in $T_{m!}$ with its counterpart
labelled with $\overline{w}$ in $\overline{T_{m!}}$,for any $w \in \{1,\ldots,b-1\}^{m!}$. This process
is also sketched in Fig. 11. Because of the symmetry,it is not
essential which of both remaining free half-edges becomes the
entrance. //

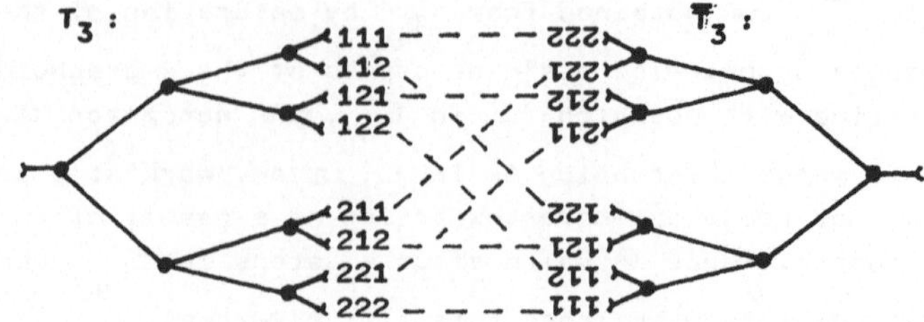

Fig. 11

Lemma 7. There is no forward automaton which searches all quasi-regular finite R-trees of degree b and always halts after the search.

This can be indirectly proved, using Lemmas 5 and 6. Assume that α is a forward automaton with m states which searches all quasi-regular finite R-trees and always halts. Without loss of generality, we suppose that a halting configuration occurs only at vertices of degree 1.

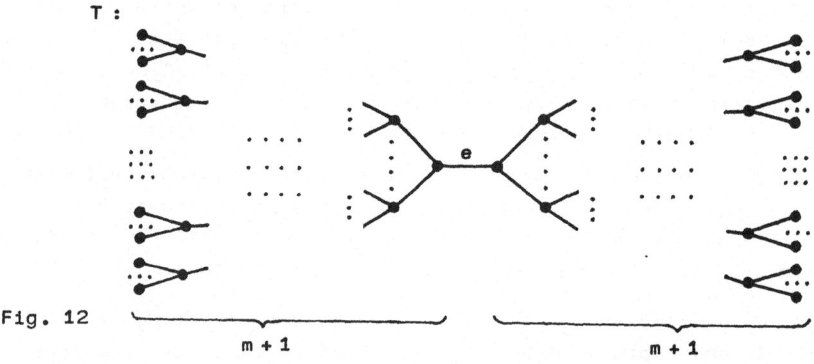

Fig. 12

Let T be the quasi-regular R-tree defined by Fig. 12 and h denote a position incident to some vertex of degree 1 in T.

Starting on position h, the forward automaton α crosses the central edge e only in cyclic states. Now we consider the quasi-regular R-ficograph $\tilde{T}$ obtained from T by inserting the corridor F from Lemma 6 into the edge e. (For the concept of edge insertion, see Section 1.10 .) If it starts with position h, the macrobehaviour of α in $\tilde{T}$, i.e., the bahaviour only considered on positions from T, is equal to its behaviour on T. Therefore, α would halt on $\tilde{T}$ after some steps of working. Then, by Lemma 5, it cannot search all quasi-regular finite R-trees, since $\tilde{T}$ contains cycles. //

The proof of Theorem 4 is obviously completed by showing the
following result.

Lemma 8. If there is a finite R-automaton which searches all
quasi-regular finite R-trees of degree b and always halts
on these labyrinths,then there is a forward automaton with
these properties.

To show this lemma,we assume that α is a finite R-automaton
having m states and searching any quasi-regular finite R-tree
of degree b and halting finally. Without loss of generality,let
α only halt at vertices of degree 1. Moreover,we can suppose
that α starting on some position first moves forward (i.e. with
output symbols $\neq 0,\S$) until it reaches a vertex of degree 1 at
the first time.

Let the vertex substitution σ on the class of all quasi-regular finite R-trees of degree b replace the stars of vertices of
degree 1 by quasi-regular complete R-trees of depth m+1, whereas
all the other stars remain unchanged. This means,

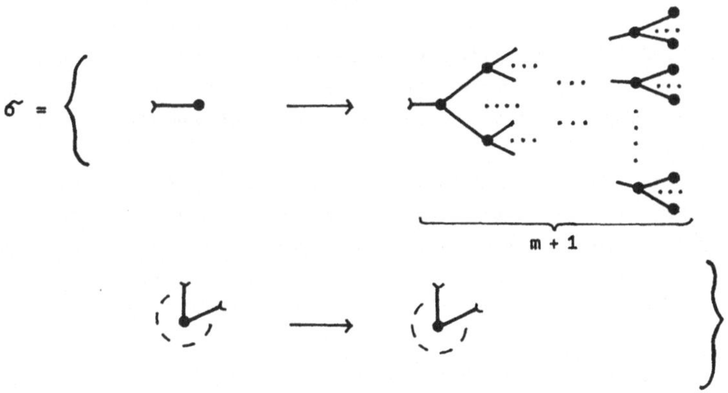

$$\sigma = \left\{ \quad \longmapsto \bullet \quad \longrightarrow \quad \underbrace{}_{m + 1} \right.$$

$$\left. \quad \longrightarrow \quad \right\}$$

Now we consider a finite automaton α' which,working on some
quasi-regular finite R-tree T, simulates the behaviour of α
working on $\sigma(T)$ in a special way.

More precisely, α' has the same set of states as α ,and for
configurations (h,a) and (h_1,a_1) of α' let

$$(h,a) \quad \vdash^{\alpha'} \quad (h_1,a_1)$$

if and only if there is a simple path

$$w_{(h,a)} = (h_0{=}h , v_0 , h_0' , h_1 , v_1 , h_1' , \ldots , v_{m+1} , h_{m+1})$$

in $\sigma(T)$ such that the following holds.
Starting from configuration (h,a) in $\sigma(T)$, α leaves the

(m+1)-neighbourhood of v_0 = ver(h) through the half-edge h_{m+1}, and a_1 is the state in which it enters the position h_1 at the last time before it leaves this neighbourhood.

If there is not such a path $w_{(h,a)}$ in $\sigma(T)$, let α' halt on position h. Then, by our supposition, deg (ver(h)) = 1.

α' can be defined as a forward automaton. This follows, since the sequence of states in which the positions $h_1, h_2, \ldots, h_{m+1}$ are entered at the first time becomes cyclic. Hence, after it has left the (m+1)-neighbourhood of v_0, the automaton α cannot return to v_0 before it has reached a vertex of degree 1 in $\sigma(T)$.

α' searches T and halts finally, since α does this on $\sigma(T)$.
// (Lemma 8 and Theorem 4)

<u>Corollary 4.</u> There is no finite C-automaton which searches all
        strongly normed quasi-regular finite 2D trees of degree 4
        and always halts finally.

This is shown by a successive reduction of labyrinth problems.

If the strongly normed quasi-regular finite 2D trees would be searchable by a halting finite automaton, then the normed quasi-regular finite 2D trees, too. This follows from the proof of Lemma 1.13 which can obviously be generalized to halting automata.

Analogously, by Lemma 1.16 , then we would obtain a halting finite automaton α which searches all quasi-regular finite 2D trees of degree 4.

By a simple simulation technique, we could now construct a finite R-automaton which searches all quasi-regular finite R-trees of degree 4 and always halts. This would contradict Theorem 4.

To show this, let T be a quasi-regular finite R-tree of degree 4 and h be a half-edge of T. There is exactly one compass system c_h on the tree T_0 underlying T such that c_h = north , the rotation system determined by c_h is equal to that of T, and that the C-tree T_h given by T_0 together with the compass system c_h is a 2D graph. The latter especially means that $c_h(h_1) = \overline{c_h(h_2)}$ for any edge $\{ h_1 , h_2 \}$ of T_0 .

Moreover, a finite R-automaton starting on h can determine the corresponding compass direction of any half-edge incident to its current position in any step. Therefore, it can simulate the work of α on T_h. //

Remark that the quasi-regularity of degree 4 is essentially used in the above proof. Using the vertex substitution σ defined in Fig. 1.24 , the degree 4 in Corollary 4 can be replaced by the degree 3, however.

Trivially,also the **normed finite 2D** trees and the finite 2D
trees cannot be searched by halting finite automata,since they
form supersets of the labyrinth type considered in Corollary 4.

Using the relationship between 2D ficographs and finite
2-dimensional mazes , as given in Section 1.4 , especially by
Lemma 1.7 , we obtain

<u>Corollary 5.</u>   There is no halting finite automaton which searches
all simply connected finite 2-dimensional mazes. //

We close this section with some remarks on further related
results. They are more detailed treated in /L.BulHe/.

One easily shows that the finite R-trees of a degree bound
$b \geqslant 3$ can be searched by a halting 1-pushdown automaton. It is not
known whether a halting 1-counter automaton can do this task or
whether it can search all finite 2D trees , cf. Problem 9 .

By Theorem 2.7 , there is a halting 1-counter automaton searching any normed 2D ficograph. Another labyrinth type which can be
searched by a halting 1-counter automaton is formed by the centred
R-trees of a degree bound $b \geqslant 3$. A finite tree is said to be
<u>centred</u> if there is a vertex v_0 such that all vertices of degree
1 have the same distance from v_0.

Using reductions by edge insertions,from Theorem 4 one obtains
that there is no halting finite automaton which searches all
quasi-regular centred R-trees of some degree $b \geqslant 3$ and all strongly
normed quasi-regular centred 2D trees of degree 4 , respectively.

<u>Hints & Sources.</u>   Theorem 3 was first published by J. Mylopoulos
/S.My/. The problem whether all finite trees of a certain type
can be searched by a halting finite automaton had been open for
some years ; see Problem 6 in the book containing /L.He85/. It was
solved in /L.BulHe/. The basic idea of the proof of Theorem 4 is
due to M. Bull.

3.3. 2D traps for finite automata

Now we are going to deal with the construction of finite 2D
traps for corresponding finite C-automata. This requires more
effort than the constructions of Section 3.1 concerning R-automata.
Like in the proof of Theorem 2,the given automaton is put into a
certain test labyrinth , and the trap is obtained by a suitable

modification of this **labyrinth according** to the behaviour of the
automaton in it. In contrast to that proof,the test labyrinth is
finite now,but it depends on the automaton,and the corresponding
modifications are more complicated.

Remember that,by Theorem 2.10,the cardinality of the faces of
2D traps or their rotation number cannot be universally bounded.
The number of faces can be bounded by three,as will be shown.

<u>Theorem 5.</u> To any finite C-automaton α with a direction set
containing D_2,there is a trap $(\,L\,,h\,)$ such that L is a
2D ficograph having at most three faces and $2^{O(\sqrt{m \cdot \ln(m)}\,)}$
vertices,where m is the number of internal states of α.

We start the construction with the C-ficograph L given in
Fig. 13. Here $e_1 = (h_1,h_1')$ and $e_2 = (h_2,h_2')$ denote directed
edges,and $c_L(h_1) =$ north , $c_L(h_1') =$ west , $c_L(h_2) =$ south ,
$c_L(h_2') =$ east. u_0 is the only vertex .

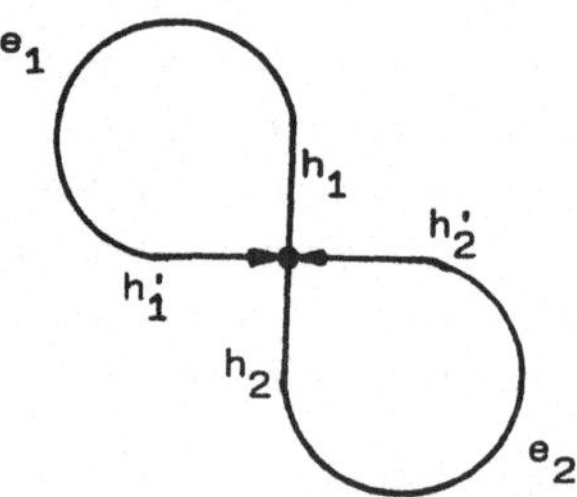

Fig. 13

By drawings we define some 2D corridors , namely

$N_r =$
1
2
$N_1 =$
1
2
$S_r =$
2
1
$S_1 =$
2
1
$C_1 =$
1
2
$C_2 =$
2
1

where the 1 s mark the entrances and the 2 s the exits.

Let α denote the given finite C-automaton with the direction set D_2 and with m internal states.

By Proposition 1.3 , there is a number

$$k^* \leq 2^{O(\sqrt{m \cdot \ln(m)})}$$

such that

$$F^{k^*} \equiv_\alpha F^{i \cdot k^*}$$

for all $i \in \mathbb{N}^+$ and any corridor $F \in \left\{ N_r , N_1 , S_r , S_1 \right\}$.

The <u>test</u> <u>labyrinth</u> for α is the C-ficograph $\Gamma = \varphi(L)$ obtained from L by the edge insertion

$$\varphi = \left\{ e_1 \longrightarrow N_r^{k^*} \cdot N_1^{k^*} \cdot C_1 \, , \, e_2 \longrightarrow S_r^{k^*} \cdot S_1^{k^*} \cdot C_2 \right\},$$

see Fig. 14. By E_1 and E_2, we denote the directed hyperedges arisen from e_1 and e_2, respectively.

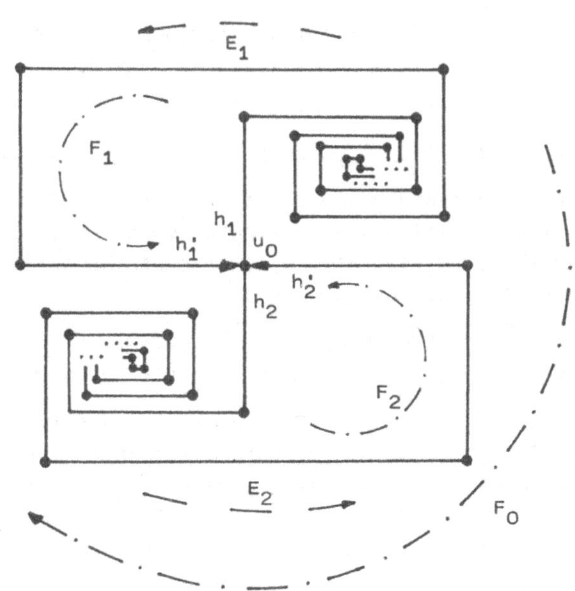

Fig. 14

In the following , we do not explicitly distinguish between corridors and directed hyperedges. Paths and , correspondingly , faces in 2D ficographs are simply denoted by the sequence (concatenation) of the traversed directed (hyper-) edges. For a corridor, a hyperedge or a path F, F^{-1} denotes the inverse one which is obtained by interchanging entrance and exit. So we have $S_r = N_1^{-1}$, and $S_1 = N_r^{-1}$. For $k < 0$, F^k is defined to be the power $(F^{-1})^{-k}$. $F^0 = \Lambda$ is the empty path. The alphabet of directed hyperedges ,

$$\overline{X} = \left\{ E_1 , E_2 , E_1^{-1} , E_2^{-1} \right\} ,$$

will be used later on.

Γ possesses just $16 \cdot k^{*} + 7$ vertices and the following tree faces,

$$F_1 = E_1 ,$$
$$F_2 = E_2 ,$$
$$F_3 = E_1^{-1} \cdot E_2^{-1} .$$

The closed rotation indices of the faces can easily be computed.

$$\overline{rin}(F_1) = rin(N_r^{k^{*}} \cdot N_1^{k^{*}} \cdot C_1) + 1 = -4 k^{*} + 4 k^{*} + 3 + 1 = 4 ,$$

$$\overline{rin}(F_2) = rin(S_r^{k^{*}} \cdot S_1^{k^{*}} \cdot C_2) + 1 = -4 k^{*} + 4 k^{*} + 3 + 1 = 4 ,$$

$$\overline{rin}(F_0) = rin(E_1^{-1}) + 1 + rin(E_2^{-1}) + 1$$
$$= 2 - rin(E_1) - rin(E_2) = 2 - 3 - 3 = -4 .$$

From Proposition 1.2 it follows that Γ is a 2D ficograph.

Let $\overline{B} = ((h_{i_0}, a_{i_0}) , (h_{i_1}, a_{i_1}) , \ldots)$ denote the macro-behaviour of the automaton α starting with position $h_{i_0} = h_1$ in the labyrinth $\Gamma = \varphi (L)$,cf. Section 1.10 .
It is either finite with a length $\leqslant 4 m$, or it is ultimately cyclic ,where the sum of the lengths of the preperiodic and of the periodic part is $\leqslant 4 m$.

<u>Case 0.</u> $\overline{B}$ is finite ,

i.e., $\overline{B} = ((h_{i_0}, a_{i_0}) , \ldots , (h_{i_1}, a_{i_1}))$ with $l < 4 m$.

This means that the $(l+1)$ st macrostep consists of infinitely many steps of the automaton. It enters a certain macroedge in this macrostep and never returns to vertex u_0 . In this case the trap can be constructed as a 2D tree similarly to the proof of Lemma 5.

Let $\overline{L}_0$ be obtained from Γ by cutting the macroedges E_1 and E_2 at the half-edges h_1' and h_2' ,respectively,as shown in Fig. 15.
The half-edge h_i'' is said to be the <u>counterpart</u> of h_i' ,and conversely,for $i = 1,2$. We shall apply this notation to the images of h_i' and h_i' in isomorphic copies of $\overline{L}_0$, too. For $k \in \mathbb{N}$, Γ_{k+1} is obtained from L_k by gluing together all free half-edges with their counterparts in mutually disjoint further isomorphic copies of $\overline{L}_0$. Finally,let $\overline{L}_k$ be obtained from Γ_k by <u>saturation</u> , this means by interlinking all free half-edges with mutually

disjoint copies of the **2D graphoids** $\rightarrowtail\!\bullet$ and $\bullet\!\multimap$, respectively,such that a 2D ficograph arises.

Starting with position h_1 in $\overline{\overline{L}}_{1+1}$, in any step of its working,the automaton α receives the same input information as in
the test labyrinth L . Therefore,it never reaches a vertex of
degree 1 , and ($\overline{\overline{L}}_{1+1}$, h_1) is a 2D trap for α.

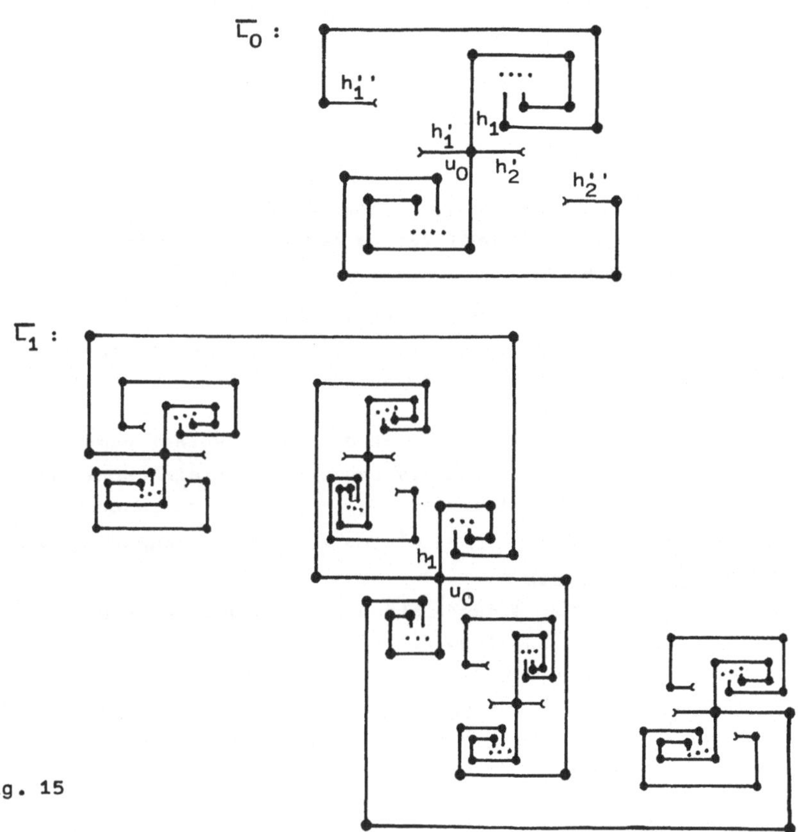

Fig. 15

As one easily proves , L_k consists of $1 + 4\cdot\sum_{i=0}^{k-1} 3^{\,i}$ copies
of L_0 and has $4\cdot 3^k$ free half-edges , for $k \in \mathbb{N}^+$.

From the supposed form of $\mathcal{B}$ it follows that the automaton α
enters at most $1+1$ copies of L_0 in the trap ($\overline{\overline{L}}_{1+1}$, h_1).
Therefore,we remove all L_0 - copies which are not entered by α
and saturate the arising free half-edges by suitable 1-vertex
graphoids. The 2D tree $\widetilde{L}$ obtained from $\overline{\overline{L}}_{1+1}$ in this way has

at most $(1+1) \cdot (16 k^* + 7 + 3) \leq 4 m \cdot (16 k^* + 10)$
$$\leq 2^{O(\sqrt{m \cdot \ln(m)})} \quad \text{vertices},$$
and $(\tilde{L}, h_1)$ is an α-trap, too.

Now we suppose that $\bar{B}$ is infinite. Then it is ultimately cyclic,
$$(h_{i_j}, a_{i_j}) = (h_{i_{j+1}}, a_{i_{j+1}}) \quad \text{for } j \geq l_0,$$
where $l_0 + 1 \leq 4 m$.

Let $\bar{w} \in \bar{X}^* \cup \bar{X}^{\mathbb{N}}$ (remember that $\bar{X} = \{E_1, E_2, E_1^{-1}, E_2^{-1}\}$) denote the <u>macropath</u> corresponding to $\bar{B}$. This is the sequence of directed hyperedges traversed by α. More precisely, let a macro-step of α (which starts on h_1 in L) yield a letter in $\bar{w}$ iff α completely traverses the corresponding directed hyperedge in this macrostep. The macropath can be represented in the form
$$(\ast) \qquad \bar{w} = w_0 \cdot w^{\infty},$$
where $w_0, w \in \bar{X}^*$ and $\mathrm{len}(w_0) + \mathrm{len}(w) \leq 4 m$.

We shall construct a so-called mp-trap for the macropath $\bar{w}$. It will be built by interlinking suitable α-equivalent copies of L_0.

A 2D ficographoid is said to be an <u>α-equivalent copy</u> of L_0 if it can be obtained from L_0 by replacing certain corridors by α-equivalent ones. For instance, $N_r^{k^*}$ could be replaced by $N_r^{i \cdot k^*}$, or $S_1^{k^*}$ by $S_1^{j \cdot k^*}$, where $i, j \in \mathbb{N}^+$. Of course, any isomorphic copy is an α-equivalent copy.

By an <u>mp-trap</u> for a macropath w', we understand a pair $(\hat{L}, \hat{h})$ satisfying the following conditions.

- $\hat{L}$ is an open 2D ficographoid built of some mutually disjoint α-equivalent copies of L_0. These graphoids are glued together by interlinking certain free half-edges with their counterparts (in other copies).

- Let be given a finite automaton $\alpha_{w'}$ traversing the macropath w' in L if it starts with position h_1, and let $\alpha_{w'}$ move in such a way that it completely traverses any hyperedge entered in some macrostep. Then, starting with position $\hat{h}$ in $\hat{L}$, $\alpha_{w'}$ never leaves this graphoid.

Assume that we have shown how to construct an mp-trap $(\hat{L}, \hat{h})$ for the macropath w^{∞}. By gluing together all free half-edges of $\hat{L}$ with their counterparts in mutually disjoint further isomorphic copies of L_0, we obtain an open 2D ficographoid $\hat{L}_1$ which is not leaved by any finite automaton with the macropath w^{∞} if it starts on $\hat{h}$. Indeed, in a macrostep in which it only enters but not traverses a hyperedge,such an automaton can at most reach one of the $\overline{L}_0$-copies added to $\hat{L}$, but not leave it.

By a backtracking argument like that used in Section 3.1 and by adding further copies of $\overline{L}_0$ if necessary,we obtain an open trap $(\hat{L}_2,h_0)$ catching the automaton α with the macropath $\overline{w}$. Finally,the free half-edges of $\hat{L}_2$ can be saturated by suitable 1-vertex graphoids,and we have a finite 2D ficograph $\hat{L}_3$ such that $(\hat{L}_3,h_0)$ is an α-trap.

Since $\hat{L}_3$ is obtained from $\hat{L}$ by a successive adding of trees, it has as many faces as the saturation of $\hat{L}$. Moreover,if the number of vertices of $\hat{L}$ is bounded by $2^{O(\sqrt{m\cdot\ln(m)})}$, then this holds for $\hat{L}_3$, too.

Therefore,to prove Theorem 5 it suffices to construct an mp-trap for the periodic part w^∞ of the macropath $\overline{w}$ of the automaton α. This problem can be reduced again.

Let $\hat{w}$ be obtained from w by application of the reduction rules from
$$\left\{ \ E_i\,E_i^{-1} \longrightarrow \Lambda \ , \ E_i^{-1}\,E_i \longrightarrow \Lambda \ : \ i = 1\,,\,2 \ \right\}$$
as often as possible. This corresponds to the reduction modulo trees , cf. Section 3.1 .

Without loss of generality,we suppose that either

(1) $\qquad \hat{w} = \Lambda$, or

(2) $\qquad \hat{w} = E_i^{\,n}$,with $i \in \{1,2\}$ and an integer $n \neq 0$, or

(3) $\qquad \hat{w} = E_1^{\,n_1}\,E_2^{\,n_2}\,\ldots\,E_1^{\,n_{2k-1}}\,E_2^{\,n_{2k}}$,with $k \in \mathbb{N}^+$ and

$$\text{integers } n_1\,,\,n_2\,,\,\ldots\,,\,n_{2k} \neq 0.$$

Indeed,let $\hat{w}$ contain both a letter from $\{E_1,E_1^{-1}\}$ and a letter from $\{E_2,E_2^{-1}\}$. Then by enlarging w_0 and shifting w , the representation (∗) can be modified to
$$\overline{w} = w_0'\cdot(w')^\infty$$
such that $\mathrm{len}(w_0') \leqslant \mathrm{len}(w_0\,w) \leqslant 4\,m$, $\mathrm{len}(w') = \mathrm{len}(w)$, and the reduction of w' yields a word $\hat{w}'$ of the form (3).

<u>Lemma 9.</u> To any macropath $\hat{w}^\infty$,where $\hat{w}$ satisfies (1),(2) or (3), there is an mp-trap $(\hat{L},\hat{h})$ such that the saturation of $\hat{L}$ possesses at most three faces and $2^{O(\sqrt{m\cdot\ln(m)})}$ vertices.

From such a trap for the macropath $\hat{w}$,one easily obtains an mp-trap $(\hat{\hat{L}},\hat{\hat{h}})$ for w^∞. This is possible by successive adding of at most $4\,m\cdot2^{O(\sqrt{m\cdot\ln(m)})}$ isomorphic copies of $\overline{L}_0$ to $\hat{L}$. Hence,Lemma 9 together with the previous part of the proof imply Theorem 5.

We remark that the reduction modulo trees for C-automata is

even simpler than in the R-case, since the reduction of a sequence w^∞ directly derives from the reduction of the word w if it starts with $E_1^{n_1}$ and terminates with $E_2^{n_2}$, for some $n_1, n_2 \neq 0$.

Now we prove Lemma 9 by discussing the several cases according to the form of the word $\hat{w}$.

<u>Case 1.</u> $\hat{w} = \Lambda$.

Then $(\hat{L}, \hat{h}) = (L_0, h_1)$ already satisfies all required properties, since the automaton α_Λ never changes its position.

If E is a hyperedge containing a corridor F just once and F' is a corridor α-equivalent to F, then let $E(F/F')$ denote the hyperedge obtained from E by replacing F by F'.

Given a word $\bar{p} \in \overline{X}^*$ and a hyperedge E, let

$$
\tilde{E} = \begin{cases} E & \text{if } \overline{rin}(\bar{p}) = 0, \\ E(N_r^{k^*}/N_r^{k^*(1+i)}) & \text{if } \overline{rin}(\bar{p}) = 4i > 0, \\ E(N_l^{k^*}/N_l^{k^*(1+i)}) & \text{if } \overline{rin}(\bar{p}) = 4i < 0, \end{cases}
$$

where $i \in \mathbb{N}^+$. It follows
$$
rin(\tilde{E}) = rin(E) - k^* \cdot \overline{rin}(\bar{p}).
$$
Moreover, replacing E by $\tilde{E}$ in L_0, one obtains an α-equivalent copy. By Lemma 1.14, the macrobehaviour of α, starting on some image of the half-edge h_1 in a labyrinth consisting of copies of L_0, is not changed by such a replacement.

Without loss of generality, we suppose $k^* > 2$ in the sequel.

<u>Case 2.</u> $\hat{w}^\infty = \bar{p}^\infty$, where $\bar{p} \in \overline{X}^+$, $len(\bar{p}) \leq 4m$ and $\overline{rin}(\bar{p}) \in \{4, -4\}$.

This applies if $\hat{w}$ is of the form (2), but $\bar{p} = E_1 E_2$ or $p = E_1^{-1} E_2^{-1}$ are possible, too.

Let $\bar{p} = p_0 E$, where $p_0 \in \overline{X}^*$ and $E \in \overline{X}$. We define a hyperedge q by
$$
q = \bar{p}^{k^*} \cdot p_0 \tilde{E}.
$$
Then $\overline{rin}(q) = (k^* + 1) \cdot \overline{rin}(\bar{p}) - k^* \cdot \overline{rin}(\bar{p}) = \overline{rin}(\bar{p})$.

Let $\hat{L}$ be constructed by gluing together L_0 with $(k^* + 1) \cdot len(\bar{p}) - 2$ isomorphic copies and one α-equivalent copy of itself in a way corresponding to q. More precisely, L_0 corresponds to the first letter of q, the isomorphic copies correspond to the remaining letters of $\bar{p}^{k^*} \cdot p_0$, and the α-equivalent copy corresponds to $\tilde{E}$. They are glued together by interlinking suitable free half-edges with the images of their counterparts in a cyclic way "via q", as sketched in Fig. 16.

Starting on h_1 in $\hat{L}$, for $w' = \hat{w}^\infty$, an automaton $\alpha_{w'}$ follows the macropath q^∞ in a cyclic manner and never leaves the graphoid.

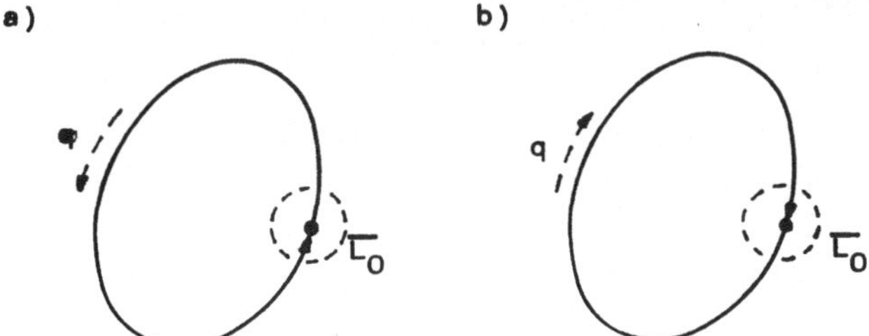

Fig. 16

The saturation of $\hat{L}$ has just two faces which (up to trees) correspond to the cycles q and q^{-1}, respectively. Especially their closed rotation indices are equal to those of q and q^{-1}. By Proposition 1.2, $\hat{L}$ is a 2D ficographoid.

It consists of $(k^{\times} + 1) \cdot \text{len}(\bar{p})$ copies of $\bar{L}_0$, thus it has $\leqslant (k^{\times} + 1) \cdot 4m = 2^{O(\sqrt{m \cdot \ln(m)})}$ vertices.

The assertion of Lemma 9 has been shown for Case 2.

We remark that the interior face of the saturation of $\hat{L}$ lies to the left of the macropath q if $\overline{\text{rin}}(\bar{p}) = 4$, but to the right otherwise (i.e. if $\overline{\text{rin}}(\bar{p}) = -4$). From this reason, Fig. 16 contains two drawings. These orientations can be changed by taking the macropath

$$q' = \bar{p}^{(k^{\times} - 2)} \cdot p_0 \tilde{E}$$

instead of q. Indeed, $\overline{\text{rin}}(q') = (k^{\times} - 1)\overline{\text{rin}}(\bar{p}) - k^{\times}\overline{\text{rin}}(\bar{p}) = -\overline{\text{rin}}(\bar{p})$.

For a concatenation $p_1 p_2$ of non-empty macropath p_1 and p_2, let $\text{prin}(p_1, p_2)$ denote the rotation index of the piece connecting the last half-edge of p_1 with the first one of p_2. For instance,

$$\text{rin}(p_1 p_2) = \text{rin}(p_1) + \text{rin}(p_2) + \text{prin}(p_1, p_2) \quad \text{and}$$
$$\overline{\text{rin}}(p_1) = \text{rin}(p_1) + \text{prin}(p_1, p_1) \quad \text{if } p_1 \text{ is closed}.$$

<u>Case 3.</u> $\hat{w}^{\infty} = (E^n p_0)^{\infty}$, where $n \geqslant 2$, $\text{len}(E^n p_0) \leqslant 4m$, $E \in \bar{X}$ and $p_0 \in \bar{X}^{+}$ such that both the first and the last letter of p_0 are different from E and E^{-1}.

This corresponds to the form (3) of $\hat{w}$, where $n_i \geqslant 2$ for some exponent n_i (and w_0 is sufficiently enlarged if necessary).

Let

$$\bar{p} = E^n p_0 ,$$
$$q_0 = E^{n-1} ,$$
$$q_1 = \tilde{E} p_0 (\bar{p})^{k^{\times} - 1} E ,$$
$$q_2 = p_0 \tilde{E} E^{n-1} p_0 (\bar{p})^{k^{\times} - 2} .$$

It follows

$$\overline{rin}(q_0) = (n-1)\cdot\overline{rin}(E) \ ,$$
$$\overline{rin}(q_1) = (2-n)\cdot\overline{rin}(E) \ ,$$
$$rin(q_2) = rin(p_0) - \overline{rin}(\overline{p}) \ .$$

Let the labyrinth $\hat{L}$ be built of L_0 and some isomorphic and two α-equivalent copies of L_0 (where E is replaced by $\tilde{E}$) such that the macropath q_0 leads from u_0 to an u_0-image w_0' and q_1 and q_2 lead from u_0' to u_0, cf. Fig. 17.

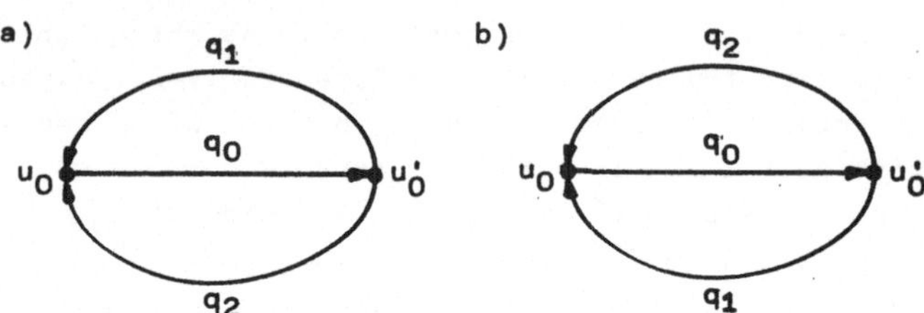

Fig. 17

$\hat{L}$ consists of $len(q_0) + len(q_1) + len(q_2)$ copies of L_0, thus it possesses at most $(2k^{*}-1)\cdot 4m\cdot 2^{O(\sqrt{m\cdot\ln(m)})} = 2^{O(\sqrt{m\cdot\ln(m)})}$ vertices.

For $w' = \hat{w}^{\infty}$, an automaton $\alpha_{w'}$ follows the macropath $(q_0 q_1 q_0 q_2)^{\infty}$ if it starts with position h_1 at u_0 in $\hat{L}$, i.e., it never leaves this labyrinth.

It remains to show that the saturation of $\hat{L}$ is a 2D ficograph having only three faces. Depending on the value of $\overline{rin}(E)$, we have to discuss two subcases.

<u>Subcase 3.1.</u> $\overline{rin}(E) = 4$, i.e. $E \in \{E_1, E_2\}$.

Then the faces of the saturation of $\hat{L}$ correspond to the cycles
$$F_1 = q_0 q_1 \ ,$$
$$F_2 = q_0^{-1} q_2^{-1} \ ,$$
$$F_3 = q_1^{-1} q_2 \ .$$

We compute the closed rotation indices of these faces.

$$\overline{rin}(F_1) = rin(q_0) + rin(q_1) + prin(q_0,q_1) + prin(q_1,q_0)$$
$$= \overline{rin}(q_0) + \overline{rin}(q_1) = \overline{rin}(E) = 4 \ ,$$

$$rin(F_2) = -\overline{rin}(q_2 q_0) = -(rin(q_2) + rin(q_0) + prin(q_2,q_0)$$
$$+ prin(q_0,q_2))$$
$$= -(rin(p_0) - \overline{rin}(\overline{p}) + (n-1)\overline{rin}(E) - prin(E,E)$$
$$+ prin(p_0,E) + prin(E,p_0))$$
$$= -(-\overline{rin}(E)) = \overline{rin}(E) = 4 \ ,$$

$$\overline{rin}\,(F_3) = -\,rin(q_1) + rin(q_2) + prin(E^{-1},p_0) + prin(p_0,E^{-1})$$

$$= (n-2)\ \overline{rin}(E) + prin(E,E) + rin(p_0) - \overline{rin}(\overline{p})$$
$$+ prin(E^{-1},p_0) + prin(p_0,E^{-1})$$

$$= -\,2\,\overline{rin}(E) + 2\,prin(E,E) - prin(p_0,E) - prin(E,p_0)$$
$$+ prin(E^{-1},\,p_0) + prin(p_0,E^{-1})$$

$$= -\,8 + 2 + 1 + 1 = -\,4.$$

Therefore, F_1 and F_2 are interior faces and F_3 is the exterior face of the saturation of $\hat{L}$; and this is a 2D ficograph by Proposition 1.2 . Fig. 17 a sketches this situation.

<u>Subcase 3.2.</u> $rin\,(E) = -\,4$, i.e. $E \in \left\{ E_1^{-1}, E_2^{-1} \right\}$.

Here we obtain the faces

$$F_1 = q_0\,q_2\ ,$$
$$F_2 = q_0^{-1}\,q_1^{-1}\ ,$$
$$F_3 = q_1\,q_2^{-1}$$

with the closed rotation indices

$$\overline{rin}(F_1) = -\,\overline{rin}(E) = 4\ ,$$
$$\overline{rin}(F_2) = -\,\overline{rin}(E) = 4\ ,$$
$$rin(F_3) = 2\,rin(E) - 2\,prin(E,E) + prin(E,p_0) + prin(p_0,E)$$
$$+ prin(E,p_0^{-1}) + prin(p_0^{-1},\,E\,)$$
$$= -\,8 + 2 + 1 + 1 = -\,4\ .$$

The corresponding situation is sketched in Fig. 17 b .

<u>Case 4.</u> $\hat{w} = E_1^{n_1} E_2^{n_2} \ \ldots\ E_1^{n_{2k-1}} E_2^{n_{2k}}$, where $len(\hat{w}) \leqslant 4\,m$
and $\left\{ n_i : 1 \leqslant i \leqslant 2k \right\} = \left\{ 1, -1 \right\}$.

This corresponds to the form (3) of $\hat{w}$ if Cases 2 and 3 do not apply.

Without loss of generality, we suppose $\left\{ n_1, n_{2k} \right\} = \left\{ 1, -1 \right\}$. Let l denote the maximal length of a segment of equal exponents in $\hat{w}^{\infty}$, and l' be the length of the following segment of equal exponents. Thus $\hat{w}^{\infty}$ can be represented as

$$\hat{w}^{\infty} = \hat{w}_0 \cdot (\ E_{i_1}^{n}\, E_{i_1+1}^{n} \cdots E_{i_1+l}^{n}\, E_{i_1+l+1}^{\overline{n}} \cdots$$
$$\cdots E_{i_1+l+l'}^{\overline{n}}\, E_{i_1+l+l'+1}^{n}\, w_1\ )^{\infty}\,,$$

where $\left\{ n,\overline{n} \right\} = \left\{ 1, -1 \right\}$, $w_1 \in X^{*}$ and $l + l' + 1 + len(w_1) \leqslant 2\,len(\hat{w}) \leqslant 8\,m$. By definition, we have

$$l' \leqslant l\ .$$

Let

$$E = E_{i_1+l+1}^{\overline{n}}$$

$$E_0 = E_{i_1+1+1'+1}{}^n \ ,$$

$$p = E_{i_1+1-1'+1}{}^n \ \cdots \ E_{i_1+1}{}^n \ ,$$

$$p_0 = E_{i_1+1+1'+1}{}^n \cdot w_1 \cdot E_{i_1}{}^n \ \cdots \ E_{i_1+1-1'}{}^n \ .$$

Then p_0 starts and terminates with the same letter

$$E_0 = E_{i_1+1+1'+1}{}^n = E_{i_1+1-1'}{}^n \ ,$$

and

$$E_{i_1+1+2}{}^{\bar n} \ \cdots \ E_{i_1+1+1'}{}^{\bar n} = p^{-1} \ .$$

Therefore,

$$\hat w^\infty = \hat w_0 \cdot E_{i_1}{}^n \ \cdots \ E_{i_1+1-1'}{}^n \cdot (\, p \, E \, p^{-1} \, p_0 \,)^\infty \ .$$

We shall construct an mp-trap with the properties required in Lemma 9 for the sequence $\bar p^\infty$, where

$$\bar p = p \, E \, p^{-1} \, p_0 \ .$$

From this one obtains such an mp-trap for the sequence $\hat w^\infty$ by backtracking via $\hat w_0 \cdot E_{i_1}{}^n \ \cdots \ E_{i_1+1-1}{}^n$, and Lemma 9 would be completely proved.

To lessen the formal effort, we assume $p \ne \Lambda$ in the sequel. Let

$$q_1 = \tilde E \, p^{-1} \, p_0 \, \bar p^{\,k^\times - 1} \, p \, E \ ,$$

$$q_2 = p_0 \, \bar p^{\,k^\times - 2} \, p \, \tilde E \, p^{-1} \, p_0 \ .$$

It follows

$$\overline{rin}(q_1) = k^\times \, \overline{rin}(\bar p) + \overline{rin}(E) - k^\times \, \overline{rin}(\bar p) = \overline{rin}(E) \ ,$$

$$\begin{aligned}
\overline{rin}(q_2) = {}& k^\times \, \overline{rin}(\bar p) - (\, prin(p_0,p) + rin(p \, \tilde E \, p^{-1}) \\
& \qquad + prin(p^{-1},p_0) \,) - k^\times \, \overline{rin}(\bar p) \\
& \qquad\qquad - prin(p_0,p_0) \\
= {}& - rin(E) + prin(p,E) + prin(E,p^{-1}) \\
& \qquad - prin(p_0,p) - prin(p^{-1},p_0) - prin(p_0,p_0) \ .
\end{aligned}$$

Let the C-ficograph $\hat L$ be built of $\bar L_0$, some isomorphic copies and two α-equivalent copies of $\bar L_0$ (E is replaced by $\tilde E$ in the latter ones) such that the macropath q_2 starts and terminates at vertex u_0, q_1 starts and terminates at u_0', and p leads from u_0 to u_0', cf. Fig. 18. Starting with position h_1 at vertex u_0, the automaton $\alpha_{w'}$ for $w' = \bar p^\infty$ follows the macropath $(p \, q_1 \, p^{-1} \, q_2)^\infty$. Obviously, the number of vertices of $\hat L$ is bounded by $2^{O(\sqrt{m \cdot \ln(m)}\,)}$.

<u>Subcase 4.1.</u> $\bar n = 1$, i.e. $E \in \{ E_1 , E_2 \}$.

It follows $E_0 \in \{ E_1^{-1} , E_2^{-1} \}$.

The faces of the saturation of $\hat{L}$ correspond to

$$F_1 = q_1 ,$$
$$F_2 = q_2^{-1} ,$$
$$F_3 = p\, q_1^{-1}\, p^{-1}\, q_2 .$$

The closed rotation indices are

$$\overline{rin}(F_1) = \overline{rin}(q_1) = \overline{rin}(E) = 4 ,$$
$$\overline{rin}(F_2) = -\overline{rin}(q_2) = rin(E) + 1 + 1 - 1 = 4 ,$$
$$\overline{rin}(F_3) = rin(p) - rin(q_1) - rin(p) + rin(q_2)$$
$$+ prin(p, q_1^{-1}) + prin(q_1^{-1}, p^{-1}) + prin(p^{-1}, q_2)$$
$$+ prin(q_2, p)$$
$$= -3 - 3 + 1 + 1 = -4 .$$

Therefore, $\hat{L}$ is a 2D ficographoid , cf. Fig. 18 a .

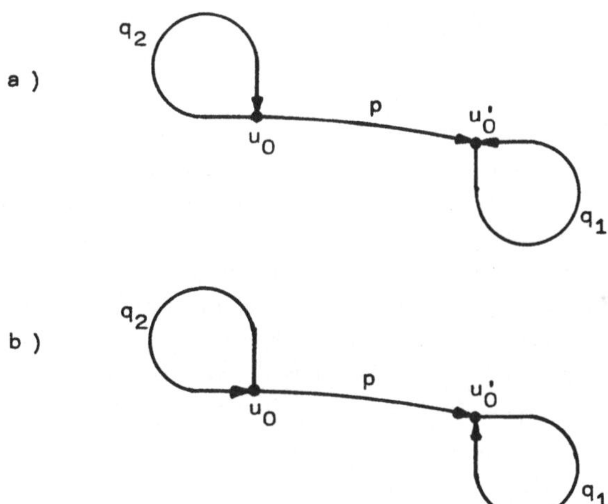

Fig. 18

Subcase 4.2. $\bar{n} = -1$, i.e. $E \in \{ E_1^{-1}, E_2^{-1} \}$.

Then $E_0 \in \{ E_1 , E_2 \}$. We obtain the faces

$$F_1 = q_1^{-1} ,$$
$$F_2 = q_2 ,$$
$$F_3 = p\, q_1\, p^{-1}\, q_2^{-1} .$$

The closed rotation indices are

$$\overline{rin}(F_1) = 4 ,$$
$$\overline{rin}(F_2) = 4 ,$$
$$\overline{rin}(F_3) = -4 .$$

So Fig. 18 b sketches the 2D ficograph $\hat{L}$.

Lemma 9 and Theorem 5 have been proved. //

Hints & Sources. Theorem 5 and its proof play a central role in labyrinth theory. We already mentioned this fact in the introduction. The proof is essentially due to L. Budach /L.Bu75,78a,b/. His traps were co-finite 2-dimensional mazes,and he considered the mastering problem. H. Müller /L.Mu79/ improved Budach's result by showing that the number of faces can be universally bounded by three. Especially he contributed the construction for Case 4. Finally,H. Antelmann /L.An/ estimated the number of vertices , cf. Section 1.7 . We have slightly modified the proof in order tó prepare the proof of certain corollaries , see the next section.

3.4. Corollaries about 2D traps

The assertion of Theorem 5 can still be improved. For this purpose,we indicate not only the starting position in a trap but also a certain goal vertex never reached by the automaton. More precisely,by a <u>strong trap</u> for some automaton α,we understand a triple (L,h,v) ,where L is a labyrinth of a corresponding type, h a half-edge and v a vertex in L such that α never reaches v if it starts with position h in L .

Let u be a vertex having both a northern and an eastern incident half-edge h_1 and h_2 in a 2D graphoid L , i.e. $c_L(h_1)$ = north and $c_L(h_2)$ = east . By the <u>NE angle</u> at vertex u ,we straightforwardly mean the angle

$$\text{ang} (h_1) = (h_1,u,h_2) .$$

It will briefly be denoted by α_u or α_{h_1} in the sequel .

<u>Lemma 10.</u> Let $j \in \{1,2,3\}$. For any finite automaton α, the trap
$(\tilde{L},\tilde{h})$ from Theorem 5 can be modified to a strong trap
$(\bar{L},\bar{h},\bar{u})$ such that $\bar{L}$ possesses all properties stated in
Theorem 5 for $\tilde{L}$ but,moreover, deg(ver($\bar{h}$)) = deg($\bar{u}$) = 4
and it holds the following condition (j') .
(1') $\alpha_{\bar{u}}$ belongs to an interior face,but $\alpha_{\bar{h}}$ to the exterior
face of $\bar{L}$.
(2') $\alpha_{\bar{u}}$ belongs to the exterior ,$\alpha_{\bar{h}}$ to an interior face.
(3') $\alpha_{\bar{u}}$ and $\alpha_{\bar{h}}$ belong to different interior faces of $\bar{L}$.

This does not follow from Theorem 5 but from our proof in the previous section which must be carefully considered once more.

If the cases O or 1 apply in the proof of Theorem 5,then $\widetilde{L}$ is a 2D tree. We can suppose that the simple path connecting $\widehat{h}$ with a vertex $\widetilde{u}$ never reached by α enters $\widetilde{u}$ from west. The additional properties (1'),(2'),(3') can be satisfied by replacing the 2D graphoid

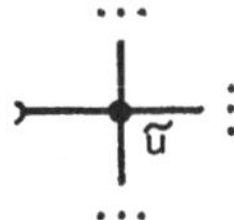

in $\widetilde{L}$ by the suitable 2D graphoid shown in Fig. 19 .

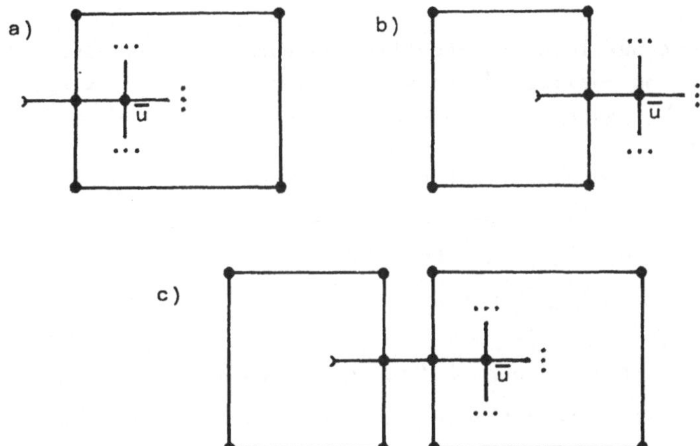

Fig. 19

If Case 2 applies,then $\widetilde{L}$ is obtained by gluing together $\widehat{L}$ from Lemma 9 with some further isomorphic copies of $\overline{L}_0$ and by saturation. $\widetilde{h}$ is obtained by backtracking from $\widehat{h}$ via w_0. Both ver($\widetilde{h}$) and a vertex $\widetilde{u}$ never reached by α are assumed to be images of vertex u_0 in $\overline{L}_0$-copies. Moreover,we suppose that $\widetilde{u}$ does not belong to a cycle in $\widetilde{L}$. The angles α_h and α_u can lie on the left-hand side or on the right-hand side of the closed macropaths q and q',cf. Fig. 16 .

If the two angles lie on different sides of the macropath q or q',then Properties (1') and (2') can be fulfilled by taking both q and q' for the construction of $\widehat{L}$ in Lemma 9. For the property (3'),we take the strong trap satisfying (2') and having only two faces,and add a further cycle surrounding $\overline{u}$ as sketched in Fig. 19 a .

Now let $\alpha_{\tilde{u}}$ and $\alpha_{\tilde{h}}$ lie on the same side of q and q'. For the properties (1') and (2'),we take the strong trap $(\tilde{L},\tilde{h},\tilde{u})$ with two faces such that $\alpha_{\tilde{h}}$ and $\alpha_{\tilde{u}}$ belong to the exterior face and add a face like shown in Fig. 19 a and b, respectively. To satisfy (3'),we modify the corresponding strong trap,where $\alpha_{\tilde{h}}$ and $\alpha_{\tilde{u}}$ belong to the interior face,according to Fig. 19 a.

Finally,we consider the cases 3 and 4 from the proof of Theorem 5 or Lemma 9. In these cases,both on the left-hand side and on the right-hand side of the macropaths q_1 and q_2 in the mp-trap $(\hat{L},\hat{h})$,there are u_0-images never reached by the automaton $\alpha_{w'}$, for $w' = \bar{p}^{\infty}$,if it starts on $\hat{h}$. Indeed,otherwise the way of $\alpha_{w'}$ in $\hat{L}$ or, equivalently, in $\hat{L}$ would be face-following,i.e. $\hat{w}^{\infty} \in \{E_1^{\infty}, E_2^{\infty}, (E_1^{-1} E_2^{-1})^{\infty}, (E_2^{-1} E_1^{-1})^{\infty}\}$, and Case 2 would apply.

Without loss of generality,we suppose $k^{*} > m$ and shift the starting position $\hat{h}$ from h_1 at vertex u_0 to h_1-images in suitable $\bar{L}_0$-copies corresponding to the macropath q_1 and q_2,respectively.

So we obtain a 2D ficograph $\tilde{L}'$ with only three faces and $2^{O(\sqrt{m \cdot \ln(m)})}$ vertices and with half-edges $\tilde{h}_1, \tilde{h}_2$ and vertices $\tilde{u}_1, \tilde{u}_2, \tilde{u}_1', \tilde{u}_2'$ such that

- $(\tilde{L}', \tilde{h}_i, \tilde{u}_j)$ and $(\tilde{L}', \tilde{h}_i, \tilde{u}_j')$ are strong traps for α if $i,j \in \{1,2\}$;
- $\deg(\tilde{u}_i) = \deg(\tilde{u}_i') = \deg(\operatorname{ver}(\tilde{h}_i)) = 4$, and the simple paths connecting $\tilde{u}_i, \tilde{u}_i'$ or $\tilde{h}_i$ with the cycles of L' meet these cycles at q_i, for $i = 1,2$;
- $\alpha_{\tilde{u}_1}$ and $\alpha_{\tilde{h}_1}$ lie on the same side of q_1 as $\alpha_{\tilde{u}_2}$ and $\alpha_{\tilde{h}_2}$ do,with respect to q_2, but $\alpha_{\tilde{u}_1'}$ and $\alpha_{\tilde{u}_2'}$ lie on the other side.

Fig. 20 illustrates this situation.

Now it is immediately clear how to satisfy the conditions (1'), (2') or (3'). //

<u>Corollary 6.</u> Let $j \in \{1,2,3\}$ and α be a finite C-automaton with a direction set containing D_2 and with at most m states. Then there is a strong trap (L,h,v) for α such that L is a 2D ficograph with at most three faces and $2^{O(\sqrt{m \cdot \ln(m)})}$ vertices, $\deg(v) = \deg(\operatorname{ver}(h)) = 1$, and it holds the following condition (j).

(1)   The vertex v touches an interior face , but ver(h)
      touches the exterior face of L.
(2)   The vertex v touches the exterior face , but ver(h) an
      interior face of L.
(3)   The vertex v and ver(h) touch different interior
      faces of L.

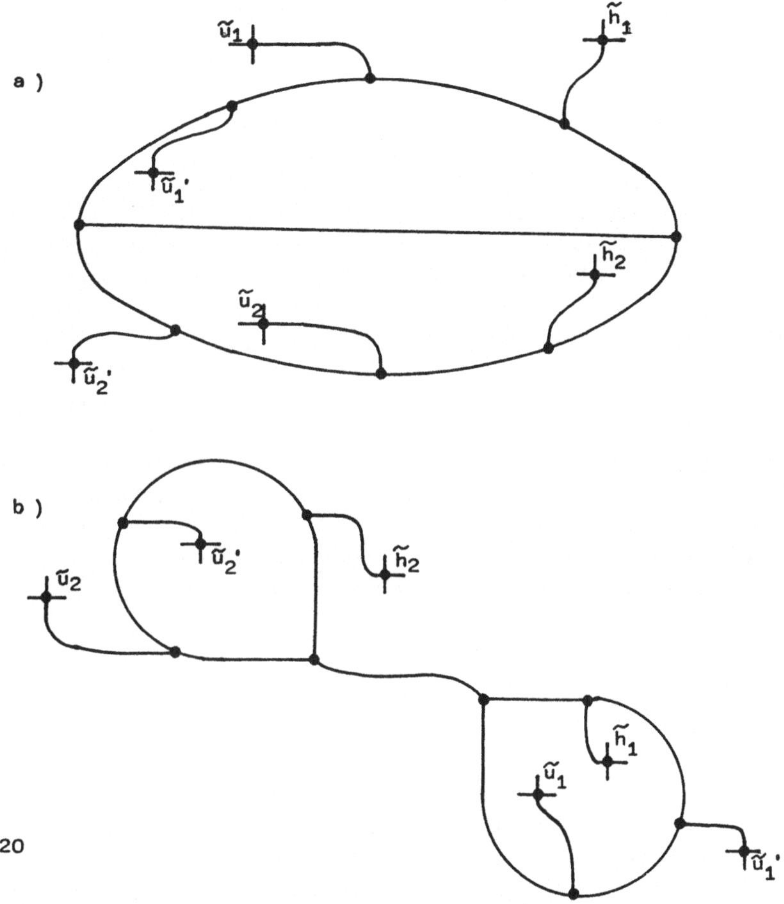

Fig. 20

Corollary 6 follows from Lemma 10 by a suitable vertex substitution , cf. Section 1.8 .

On the class of all 2D ficographs , let the vertex substitution $\tilde{\sigma}$ be defined in the following way.

$$\sigma: \left\{ \quad \begin{array}{c} h_4 \;\vdash\!\!\!\!\mid^{h_1}\!\!\!\!\dashv\, h_2 \\ h_3 \end{array} \quad \longrightarrow \quad h_4' \;\vdash\!\!\!\!\mid^{h_1'} \;\; \mid^{h_0} \; h_3' \;\dashv\, h_2' \right.$$

identical transformations for vertices
of degrees 3 .

As defined in Section 1.8 , let $\alpha^{(\sigma)}$ be the finite automaton
which, working on a 2D ficograph L , simulates the behaviour of the
given automaton α working on $\sigma(L)$, where the initial state of
$\alpha^{(\sigma)}$ corresponds to the starting position h_0 in the hypervertices of degree four. $\alpha^{(\sigma)}$ has $O(m)$ internal states.

We consider a strong trap $(\bar{L}, \bar{h}, \bar{u})$ for $\alpha^{(\sigma)}$ according to
Lemma 10. If $L = \sigma(\bar{L})$, h is the isomorphic image of h_0 in the
hypervertex replacing $\text{ver}(\bar{h})$ in $\bar{L}$, and v is the image of
$\text{ver}(h_0)$ in the hypervertex replacing $\bar{u}$, then (L,h,v) possesses
all properties required in Corollary 6. //

Using (the proofs of) Lemmas 13 and 16, one sees that the trap
labyrinth L from Corollary 6 can be constructed as a strongly
normed 2D ficograph. Moreover, we shall show that the embedding
of ver(h) in Condition (1) and of ver(v) in Condition (2) can be
assumed to lie outside the box generated by the embeddings of the
other vertices.

For a (finite) set of points in the n-dimensional space ,
$S \subseteq \mathbf{R}^n$, let the <u>box</u> of S be the smallest n-dimensional rectangle
containing S , i.e., if

$$x_{i,\min} = \min\{x_i: \text{ there is a point } (x_1,\ldots,x_i,\ldots,x_n) \epsilon S\},$$

$$x_{i,\max} = \max\{x_i: \text{ there is a point } (x_1,\ldots,x_i,\ldots,x_n) \epsilon S\},$$

for $1 \le i \le n$, then

$$\text{box}(S) = \underset{i=1}{\overset{n}{\bigtimes}} \; [x_{i,\min} , x_{i,\max}] .$$

<u>Corollary 7.</u> Let $j \epsilon \{1,2,3\}$ and α be a finite automaton as in
Corollary 6. There is a strong trap (L,h,v) possessing all
properties given in Corollary 6 but , moreover, having a
strongly normed 2D embedding P of L such that
- $P(\text{ver}(h)) \notin \text{box}(P(V_L \smallsetminus \{\text{ver}(h)\}))$ if $j = 1$, and
- $P(v) \notin \text{box}(P(V_L \smallsetminus \{v\}))$ if $j = 2$.

To show this, we first modify the strong trap (L,h,v) from
Corollary 6. Suppose that both h and the half-edge incident to
vertex v are southern half-edges. For $j = 1$, we replace the star

of ver(h) by a 2D graphoid containing a new cycle C,as shown in
Fig. 21 a ; for j = 2,the star of vertex v is replaced according to
Fig. 21 b .

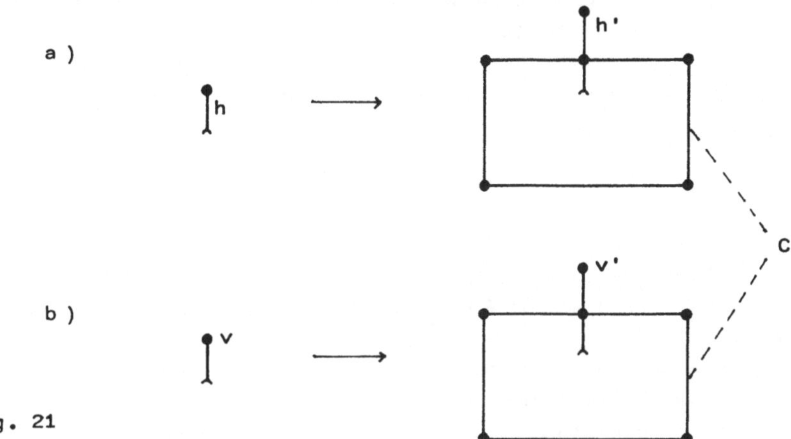

Fig. 21

Using Proposition 1.2 , one easily shows that,in both cases ,
a 2D ficograph L' with four faces arises,where ver(h') and vertex v', respectively,touch the exterior face. Moreover,(L',h',v)
and (L',h,v'),respectively,are strong traps for α.

Applying the constructions from the proofs of Lemmas 13 and 16
to these traps , strong α-traps (L'',h'',v'') with four faces and
$2^{O(\sqrt{m \cdot \ln(m)})}$ vertices are built,such that the conditions (1)
or (2) hold. By removing the rectangular shaped cycles arisen
from cycle C in these traps (except the vertex connecting h'' or
v'' with the remaining part of L''),we obtain strong traps possessing all properties stated in Corollary 7. //

There is a nice relationship between size estimations of traps
for finite automata and lower bounds of the space complexity of
Turing tape automata searching the corresponding type of labyrinths.

<u>Lemma 11.</u> Assume that $\mathcal{L}$ is a type of labyrinths such that,for
any finite automaton of the corresponding kind with at most
m states , there is a trap (L,h) , where $L \in \mathcal{L}$ and L possesses
only f(m) vertices,for some function f: $\mathbb{N}^+ \longrightarrow \mathbb{N}^+$.
Let g and s be functions of $\mathbb{N}^+$ into itself with
$$f \circ g(n) \leqslant n \quad \text{and} \quad s(n) \leqslant s(n+1) \quad \text{for any } n \in \mathbb{N}^+ .$$
If there is an s(n) space-bounded Turing tape automaton
searching all $L \in \mathcal{L}$,then $\log \circ g(n) = O(s(n))$.

For the proof we assume that $\alpha = (X, Y, A, X_T, \delta, \lambda, a_0)$ is an $s(n)$ space-bounded Turing tape automaton searching all labyrinths $L \in \mathcal{L}$, cf. Section 1.5 .

For $b = \mathrm{card}(X_T) \cdot \mathrm{card}(A)$ and any $n \in \mathbb{N}$, let α_n be a finite automaton simulating α as long as it uses not more than $\log_b(g(n)) - 1$ cells of its worktape . If α would use a further cell , α_n may halt. Then α_n can be assumed to be a finite automaton with $b^{\log_b(g(n)) - 1} \leqslant g(n)$ internal states. Thus there is a trap (L_n, h_n) for α_n, where L_n has at most $f(g(n)) \leqslant n$ vertices.

If $\log \circ g(n) = O(s(n))$ does not hold, then for any constant $k \in \mathbb{N}^+$, there are infinitely many n satisfying $k \cdot s(n) < \log \circ g(n)$. It follows $s(n') \leqslant s(n) \leqslant \log_b(g(n)) - 1$ for all these n and $n' \leqslant n$.

But for such n' and n, the automata α_n on labyrinths $L \in \mathcal{L}$ with at most n' vertices completely simulate the working of the Turing tape automaton α, i.e., they completely search L. This is a contradiction to the existence of a trap (L_n, h_n) for α_n with $n' \leqslant f(g(n)) \leqslant n$ vertices, for some n'. //

<u>Theorem 6.</u> Let $s: \mathbb{N}^+ \longrightarrow \mathbb{N}^+$. If there is an $s(n)$ space-bounded Turing tape automaton searching all strongly normed 2D fico-graphs with at most three faces , then
$$\log \circ \log(n) = O(s(n)).$$

This follows from Corollary 7 and Lemma 11 for $f(m) = 2^{k \cdot \sqrt{m \cdot \ln(m)}}$ and $g(n) = \frac{1}{l} \cdot \log(n)$ with a suitable $l \in \mathbb{N}^+$. //

<u>Hints & Sources.</u> Essentially, strongly normed strong 2D traps with the property (1) were already constructed and considered by L. Budach , H. Müller and H. Antelmann , cf. the previous section. Nevertheless, the properties (1), (2) and (3) have not yet been mentioned explicitly. Lemma 11 and Theorem 6 are from /S.HeMu/.

In this section we deal with inductive constructions of universal traps for sets of R- and C-automata. These results are
fundamental for Rabin's method of constructing traps and barrages
for cooperating systems.

Let $\mathcal{A}$ be a set of automata. A triple (L,h,v) is said to be a
underline universal strong trap for the set $\mathcal{A}$ if it is a strong trap for
any automaton α from $\mathcal{A}$. Correspondingly the concept of universal trap is defined.

We first construct universal strong traps for finite sets of
finite R-automata. Let us start with a consequence of Corollary 1
from Section 3.1 .

<u>Corollary 8.</u> To any finite R-automaton with only m internal
     states , there is a strong trap (L,h,v) satisfying the
     following properties.
     L is a plane R-ficograph of degree bound three with at most
     two faces and O(m) vertices. It holds deg(v) = 1 , and there
     is a 2-dimensional embedding P of the graph underlying L
     such that the exterior region of P touches both the point
     P(v) and the Jordan arc P(h) , the latter from the left-hand
     side.

The proof is rather simple. By Corollary 1,there is a plane
strong trap (L,h,v) with O(m) vertices,at most two faces and
deg(v) = 1. Let h' be the half-edge incident to vertex v.

If ang(h) and ang(h') belong to the same face of L,it can be
assumed to be the exterior one with respect to a suitable plane
embedding P , cf. Section 1.2 and Fig. 1.5 .

If ang(h) and ang(h') belong to different faces,let P' be a
plane embedding of L such that ang(h) belongs to the exterior
face. This embedding of L consists of a closed Jordan curve J
connected with some embedded trees , as sketched in Fig. 22 .

By transferring the tree containing vertex v to the exterior
of J , from P' we obtain a 2-dimensional embedding P of L with
the required properties. Of course,P is no plane embedding of L
in this case. //

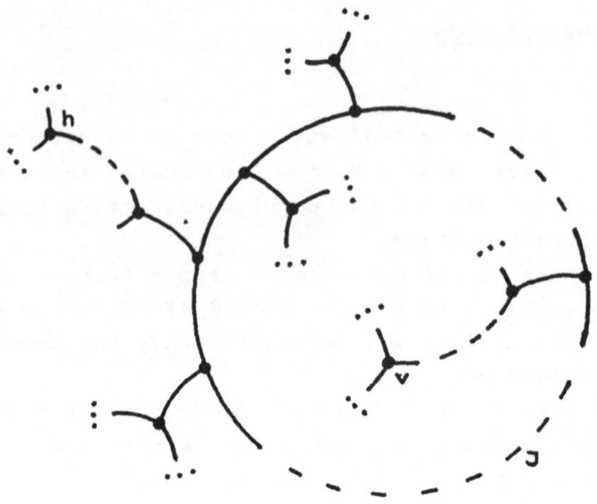

Fig. 22

<u>Theorem 7.</u> To any finite set of finite R-automata , there exists a universal strong trap (L,h,v) of degree bound three. It can be constructed in such a way that deg(v) = 1 , L is a plane R-ficograph , and there is a 2-dimensional embedding P of the ficograph underlying L , where the exterior region touches both P(v) and P(h) , the latter from the left-hand side.

This can be proved by induction on the cardinality of the given set $\mathcal{A}$. For $\mathcal{A} = \emptyset$ the assertion is trivial, for card($\mathcal{A}$) = 1 it follows from Corollary 8 .

Assume that $\mathcal{A} = \mathcal{A}_0 \cup \{\alpha\}$ and (L_0, h_0, v_0) is a universal strong trap for the set $\mathcal{A}_0$ with the properties required in Theorem 7. Let (L_0^i, h_0^i, v_0^i), i = 1,2,3 , be three mutually disjoint isomorhic copies of (L_0, h_0, v_0). By interlinking them in the manner sketched in Fig. 23 , we obtain a rotationally symmetric R-ficographoid L_0' with three free half-edges h_1', h_2', h_3' .

On the R-ficographs of degree bound 3 , we consider the vertex substitution

σ : 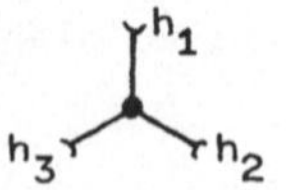 $\longrightarrow$ (L_0', h_1', h_2', h_3') ,

identical transformations for vertices of degrees ≤ 2 .

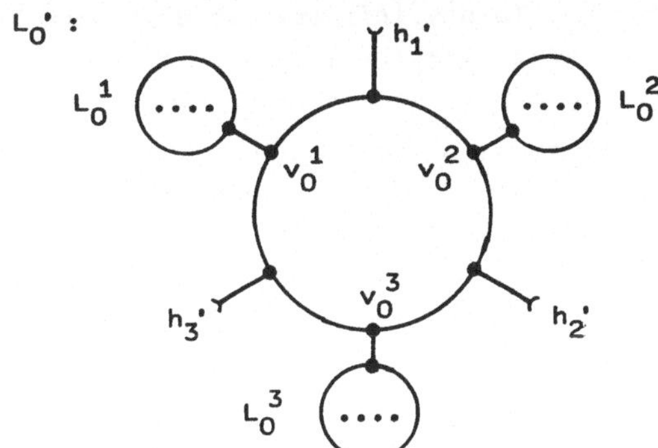

Fig. 23

According to Section 1.8 , let $\alpha_0^{(\sigma)}$ be (the short form of) an automaton which , working on a labyrinth L , simulates the behaviour of α working on $\sigma(L)$. The starting position of $\alpha_0^{(\sigma)}$ may correspond to position h_0^i within L_0' if α starts on a position corresponding to h_i at a vertex of degree 3.

We consider a strong trap (L,h,v) for $\alpha_0^{(\sigma)}$ with the properties given in Corollary 8 . One can assume that deg(ver(h)) = 3. Let h^* be the image of h_0^1 in the hypervertex replacing the star of ver(h) , where h corresponds to h_1. Then ($\sigma(L)$, h^* , v) fulfils all properties required in Theorem 7 for the universal strong trap of the set $\mathcal{A} = \mathcal{A}_0 \cup \{\alpha\}$.

Indeed, starting with position h^* corresponding to h_0^1 in L_0', the automata from $\mathcal{A}_0$ never enter the image of vertex v_0^1 in their starting hypervertex. Thus they never reach vertex v in $\sigma(L)$. But α never reaches the vertex v in $\sigma(L)$ if it starts on h^*, since $\alpha_0^{(\sigma)}$ working in L never enters v .

The 2-dimensional embedding P of $\sigma(L)$ with the required properties is obtained in a straightforward manner from an embedding P_L of L according to Corollary 8 and an embedding P_0 of L_0 according to the hypothesis of induction. //

We remark that L can be assumed to have O(m) vertices and two faces, where m gives the number of states of α. Hence, if L_0 has n_V vertices and n_F faces , $\sigma(L)$ has O(m $\cdot$ n_V) vertices and O(m $\cdot$ n_F) faces. Therefore, if all automata from the set $\mathcal{A}$ have at most m $\geqslant$ 2 states , we obtain a universal strong trap whose R-ficograph has at most $m^{O(\,card(\mathcal{A})\,)}$ vertices and faces .

The given proof enables us to transfer Theorem 7 to infinite sets of automata and to the mastering problem.

<u>Corollary 9.</u> To any infinite set of finite R-automata, there is
a universal mastering trap (L,h) such that L is a plane
R-incograph of degree bound three.

It suffices to prove Corollary 9 for countably infinite sets
of automata. Indeed, up to renaming of internal states, there are
only countably many finite R-automata.

Now let $\mathscr{A} = \{\alpha_1, \alpha_2, \ldots\}$ be the given infinite set of
automata. For $k \in \mathbb{N}^+$, let (L_k, h_k, v_k) be a universal strong trap
for the finite set
$$\mathscr{A}_k = \{\alpha_1, \alpha_2, \ldots, \alpha_k\}$$
according to the proof of Theorem 7. This means, (L_1, h_1, v_1) is
defined according to Corollary 8, and $(L_{k+1}, h_{k+1}, v_{k+1})$ is
obtained from (L_k, h_k, v_k) by the induction step described in
the proof of Theorem 7, where $\mathscr{A}_0 = \mathscr{A}_k$ and $\alpha = \alpha_{k+1}$, for $k \in \mathbb{N}^+$.
Moreover, we suppose that the isomorphic image of L_0^1 in the
hypervertex replacing $\mathrm{ver}(h)$ in that proof is equal to L_k.

So we have $h_0 = h_1 = h_2 = \ldots$, and
$$v_k \subseteq v_{k+1} \ ,$$
$$H_k \subseteq H_{k+1} \ ,$$
$$I_k \subseteq I_{k+1} \ ,$$
$$r_k / H_k \smallsetminus H(v_k) = r_{k+1} / H_k \smallsetminus H(v_k) \ ,$$
where
$$L_k = (V_k, H_k, I_k, r_k) \quad \text{for} \ k \in \mathbb{N}^+.$$
Let
$$\bar{L} = (\bar{V}, \bar{H}, \bar{I}, \bar{r}),$$
where
$$\bar{V} = \bigcup_{k \in \mathbb{N}^+} V_k \ ,$$
$$\bar{H} = \bigcup_{k \in \mathbb{N}^+} H_k \ ,$$
$$\bar{I} = \bigcup_{k \in \mathbb{N}^+} I_k \ ,$$
$$\bar{r} = \bigcup_{k \in \mathbb{N}^+} r_k / H_k \smallsetminus H(V_k) \ .$$

$\bar{L}$ is a plane R-incograph of degree bound 3, as one easily sees.
Moreover, $(\bar{L}, h_0)$ is a mastering trap for any automaton α_k, since
(L_k, h_k, v_k) is a strong trap for α_k, $h_k = h_0$, L_k is an R-sub-
graph of $\bar{L}$, and it is connected with the remaining part of $\bar{L}$

only via the vertex v_k . //

We remark that Theorem 7 and Corollary 9 can be transferred to sets of 1-pushdown automata or $f(n)$ space-bounded Turing tape automata. This follows from (the proofs of) Corollaries 2 and 3.

By similar proof techniques,we obtain universal traps for sets of C-automata.

<u>Theorem 8.</u> Let $j \in \{1,2\}$ and $\mathcal{A}$ be a finite set of finite C-automata whose direction sets contain $\mathbf{D}_2$.

There is a universal strong trap (L,h,v) for the set $\mathcal{A}$ such that $\deg(v) = \deg(\,\text{ver}(h)\,) = 1$, and L is a 2D ficograph possessing a strongly normed 2D embedding P satisfying the following condition (j).

(1) The vertex v touches an interior face , but
$$P(\,\text{ver}(h)\,) \notin \text{box}(\,P(\,V_L \smallsetminus \{\text{ver}(h)\}\,)\,) \,.$$
(2) ver(h) touches an interior face , but
$$P(\,v\,) \notin \text{box}(\,P(\,V_L \smallsetminus \{v\}\,)\,) \,.$$

Obviously,ver(h) touches the exterior face if $j = 1$, and the vertex v touches the exterior face if $j = 2$.

If $\text{card}(\mathcal{A}) = 1$,the assertion holds by Corollary 7.

We sketch the induction step for $j = 2$, for $j = 1$ it is analogous. Let $\mathcal{A} = \mathcal{A}_0 \cup \{\alpha\}$ and (L_0, h_0, v_0) be a universal strong trap for $\mathcal{A}_0$ with all properties required in Theorem 8 for $j = 2$. Without loss of generality,one can suppose that v_0 is a northern vertex of L_0, this means $c_{L_0}(h_0') = \text{south}$ if h_0' denotes the half-edge incident to v_0 .

On the class of all (strongly normed) 2D ficographs,we consider the following vertex substitution

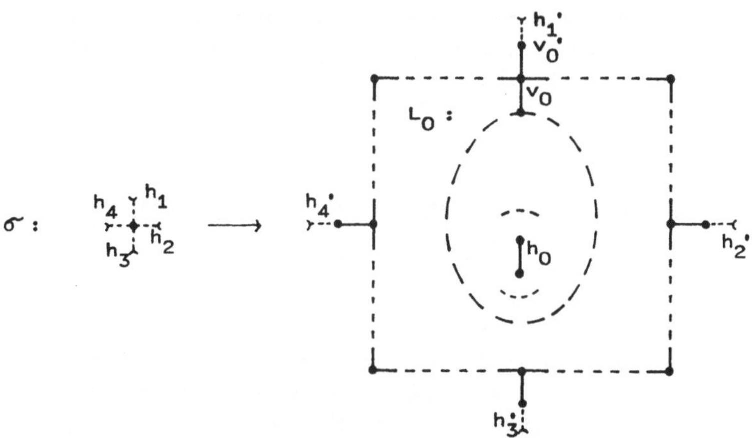

for all vertices of degrees $\geqslant 1$ (where h_i' on the right-hand
side may exist iff the half-edge h_i in the star of the left-hand
side exists, for $i = 1,2,3,4$). Obviously, if L is a strongly
normed 2D ficograph, then $\sigma(L)$, too.

Let $\alpha^{(\sigma)}$ be a finite C-automaton which, starting at some vertex v on a labyrinth L, simulates the behaviour of α working on
$\sigma(L)$ and starting at the image of the half-edge h_0 in the hypervertex corresponding to vertex v. If (L,h,v) is a strong trap
according to Corollary 7 for the automaton $\alpha^{(\sigma)}$, then we consider
the triple $(\sigma(L), h^*, v^*)$, where h^* is the image of h_0 in the
hypervertex replacing ver(h) and v^* is the image of vertex v_0' in
the hypervertex of vertex v. This is a universal strong trap for
the set $\mathcal{A} = \mathcal{A}_0 \cup \{\alpha\}$ such that the required properties (for j=2)
are fulfilled . //

<u>Corollary 10.</u> To any infinite set of finite C-automata whose
      direction sets contain D_2, there is a universal mastering
      trap (L,h) such that deg (ver(h)) = 1 and L is a strongly
      normed 2D incograph.

Using the given proof of Theorem 8 for $j = 2$, this corollary
can be shown in quite the same manner as Corollary 9. //

<u>Hints & Sources.</u> As already mentioned, universal traps play a
fundamental role in Rabin's trap construction for cooperating
systems , cf. Section 3.7 . For this purpose, they are considered
in /L.BlSa/,/L.Ro/,/L.He87a/. Theorem 8 , for j = 2 , and Corollary
10 are proved in /L.AnBuRo/. F. Hoffmann /L.Ho82/ gave another,
non-inductive proof of Theorem 8. He called the obtained traps
"universelle Fallen mit Paralleltests". The technique of constructing these universal traps is fundamental for his 1-pebble
trap construction , cf. Section 4.2 . A direct proof of Theorem 7
was given by M. Bull /L.Bul/.

3.6. Plane R-traps for plenary multihead automata

Here we deal with a slight generalization of the result by
Blum and Kozen that no two cooperating finite automata can search
every plane R-ficograph. It is based on the construction of hypervertices on which the behaviour of any automaton from some finite
set degenerates. These hypervertices will be obtained by

repeatedly applying suitable vertex substitutions,analogously to
the construction of universal traps.

As in Section 3.1 , we fix the degree bound three and shall put
the automata into cubic R-graphoids. Thus it is sufficient to
consider their autonomic behaviour corresponding to the constant
input sequence 3^{∞}. Then the sequence of current states becomes
cyclic finally. With these state cycles and the corresponding
output sequences we are going to deal in more detail now.

As defined in Section 3.2 , a state $a \in A$ of a finite R-automaton
$$\alpha = (X , Y , A , \delta , \lambda , a_0) \text{ of degree bound 3}$$
is called cyclic if
$$\delta^{*}(3^k , a) = a \quad \text{for some} \quad k \in \mathbb{N}^{+},$$
where δ^{*} denotes the natural generalization of the transition
function δ. Then under the <u>cycle</u> of state a , we understand the
cyclic permutation
$$C_a = (a = \delta^{*}(3^0,a) , \delta^{*}(3 , a) , \ldots , \delta^{*}(3^{k-1} , a)).$$
The corresponding <u>output word</u> is given by
$$w_a = y_1 y_2 \ldots y_k \in Y^{+} ,$$
where $y_i = \lambda (3 , \delta^{*}(3^{i-1} , a))$ for $1 \leq i \leq k$; $Y = \{ \S,0,1,2 \}$.

The <u>output sequence</u> of a is
$$f_a = w_a^{\infty} .$$

Given a rotationally symmetric cubic R-fragment with exactly
three free half-edges ,
$$F = (L_F , h_1 , h_2 , h_3) ,$$
let σ_F denote the vertex substitution replacing all (stars of)
vertices in cubic R-graphoids by hypervertices isomorphic to L_F ,
cf. Section 1.8 .

By $\widetilde{\alpha}^{(F)}$, we mean the finite automaton which,working on some
cubic R-graphoid.L,simulates the behaviour of α working in
$\sigma_F(L)$ and starting on the image of h_1 in the hypervertex corresponding to the starting position of α. More precisely ,
$$\widetilde{\alpha}^{(F)} = (X , Y , A , \widetilde{\delta}' , \widetilde{\lambda}' , a_0) ,$$
where $\widetilde{\delta}'$ and $\widetilde{\lambda}'$ are the restrictions of $\widetilde{\delta}$ and $\widetilde{\lambda}$ defined in
Section 1.8 for the short form $\alpha(\sigma_F)$. The addition of a new
initial state $\widetilde{a}_0$ is not necessary here,because $\widetilde{\alpha}^{(F)}$ starts on
a free half-edge of the starting hypervertex of α.

By $\widetilde{c}^{(F)}_a$, $\widetilde{w}^{(F)}_a$ and $\widetilde{f}^{(F)}_a$, we denote the cycle , the output
word and output sequence,respectively,of the state $a \in A$ in the
automaton $\widetilde{\alpha}^{(F)}$.

From the above definition it follows that

$$\widetilde{\widetilde{\alpha}^{(F_1)}}^{(F_2)} = \widetilde{\alpha}(\,\sigma_{F_1}(F_2)\,)$$

for any two rotationally symmetric cubic R-fragments F_1 and F_2.
Therefore , we also have

$$\widetilde{\widetilde{f}^{(F_1)}}^{(F_2)}_a = \widetilde{f}(\,\sigma_{F_1}(F_2)\,)_a$$

for any state a ; and corresponding equations hold for the cycles
and the output words of a . Obviously, if two states a_1 and a_2
belong to one cycle of $\widetilde{\alpha}^{(F)}$, then they also belong to one cycle
of α . In other words, from $\widetilde{c}^{(F)}_{a_1} = \widetilde{c}^{(F)}_{a_2}$ it follows that
$c_{a_1} = c_{a_2}$.

The factorizations $\widetilde{f}^{(F)}$ of control sequences $f \in Y^{\mathbb{N}}$ could
be defined quite independently on automata. We say that a control
word $w \in Y^{*}$ <u>fits</u> the fragment F if $pos_{L_F}(h_1, w')$ is defined
for any proper initial part w' of w, but $pos_{L_F}(h_1, w)$ is
undefined , i.e.,the corresponding walk leaves L_F in the last
step. Then let

$$\widetilde{w}^{(F)} = y \in \{0, 1, 2\}$$

if $pos_{L_F}(h_1, w) = hal(h_{y+1})$ in the R-ficograph L_F obtained
from L_F by saturating the free half-edges.

For a control sequence f , there is a uniquely determined representation

$$f = w_1 w_2 \ldots ,$$

where each w_i fits F ,for $i \in \mathbb{N}^{+}$, or

$$f = w_1 w_2 \ldots w_l \cdot f' ,$$

where $w_1, \ldots, w_l$ fit F ,but f' contains no initial word
fitting F . Then we define

$$\widetilde{f}^{(F)} = \widetilde{w_1}^{(F)} \widetilde{w_2}^{(F)} \ldots$$

in the first case , and

$$\widetilde{f}^{(F)} = \widetilde{w_1}^{(F)} \widetilde{w_2}^{(F)} \ldots \widetilde{w_l}^{(F)} \cdot \S^{\infty}$$

in the latter case. Obviously ,

$$(\widetilde{f_a})^{(F)} = \widetilde{f}^{(F)}_a$$

for any state $a \in A$.

Now we want to use the reduction modulo trees. Recall that ,
for control words $w, w' \in Y^{*}$, $w \longmapsto w'$ means that w' can be
obtained from w by applying a rule of the semi-Thue system $\mathcal{R}$,
see Section 3.1 . $w \overset{k}{\longmapsto} w'$ means that $w \longmapsto w_1 , w_1 \longmapsto w_2 , \ldots$
$\ldots, w_{k-1} \longmapsto w'$ for suitable words $w_1, \ldots, w_{k-1}$ and $k \in \mathbb{N}^{+}$.

$w \overset{0}{\vdash} w'$ is equivalent to $w = w'$.

A control word $w \in Y^+$ is said to be __progressive__ if $w \overset{k}{\vdash} \hat{w}$ for some $k \in \mathbb{N}$ and $\hat{w} \in \{1,2\}^+$. This is the case iff any walk according to w in the infinite cubic R-tree terminates with a position different from its starting position and reachable from this by forward moves only.

__Lemma 12.__ Any ultimately cyclic control sequence f can be represented in the form
$$f = w_0' \cdot w'^{\infty},$$
where either $w' \overset{k}{\vdash} \Lambda$, for some $k \in \mathbb{N}^+$, or w' is a progressive control word.

Indeed, if $f = w_0 \cdot w^{\infty}$ with $w_0, w \in Y^{*}$, by Lemma 4 we obtain a representation $w = w_1 w_2$, where $w_2 \neq \Lambda$ and $(w_2 w_1)^6 \overset{k}{\vdash} \hat{w}$ for some $k \in \mathbb{N}$ and $\hat{w} \in \{1,2\}^{*}$. The statement of Lemma 12 follows with $w_0' = w_0 w_1$ and $w' = (w_2 w_1)^6$. //

We shall say that a state $a \in A$ is __progressive__ if w_a is a progressive control word.

A control sequence f is said to be __degenerate__ if it can be represented in the form
$$f = w_0 \cdot w^{\infty},$$
where $w \overset{k}{\vdash} \hat{w}$ for some $k \in \mathbb{N}$ and $\hat{w} \in \{1\}^{*} \cup \{2\}^{*}$. This means that a walk controlled by f in a cubic R-ficograph L either remains within the k-neighbourhood of $\mathrm{pos}_L(h, w_0)$, namely if $\hat{w} = \Lambda$, or it remains in the k-neighbourhood of the face of $\mathrm{ang}(\mathrm{pos}_L(h, w_0))$ or of $\mathrm{ang}(\mathrm{hal} \circ \mathrm{pos}_L(h, w_0))$, where h denotes the starting position.

A cycle of α or of some $\widetilde{\alpha}^{(F)}$ is called degenerate if it contains a state whose output sequence is degenerate. Then it obviously follows that all states belonging to that cycle have degenerate output sequences. Moreover, if a cycle C_a is degenerate, then $\widetilde{C}_a^{(F)}$, too , for any fragment F of the above described kind.

__Lemma 13.__ For any cycle C_a of a finite R-automaton α, there are a state a_0 belonging to C_a and a rotationally symmetric cubic plane R-fragment F such that
   i) $\widetilde{C}^{(F)}_{a_0}$ is degenerate , and
   ii) $\sigma_F(L)$ is plane, for any cubic plane R-graphoid L.

If C_a is degenerate , the trivial R-fragment
$$F = \quad$$

fulfils the assertion of the lemma.

Now let the control sequence f_a be non-degenerate.
By Lemma 12 ,

$$f_a = w_0' \cdot w'^{\infty}$$

with a progressive control word w' , i.e. $w' \xrightarrow{k} \hat{w}$ for some $k \in \mathbb{N}$
and $\hat{w} \in \{1,2\}^+$.

We can suppose that $\text{len}(\hat{w}) \geqslant 2$ and either $\hat{w} = 1\,2$ or $\hat{w}$ starts
and terminates with the same letter $y_0 \in \{1,2\}$. For instance, if
$\hat{w}$ contains a word $y_0 y_0$ with $y_0 \in \{1,2\}$, then we obtain representations $\hat{w} = \hat{w}_1 \hat{w}_2$ and $w' = w_1' w_2'$ such that $w_i' \xrightarrow{k_1} \hat{w}_1$, for
$i = 1,2$ and $k_1 + k_2 = k$, and $\hat{w}_2 \hat{w}_1$ starts and terminates with y_0 .
Then we have $f_a = w' w_1' \cdot (w_2' w_1')^{\infty}$.

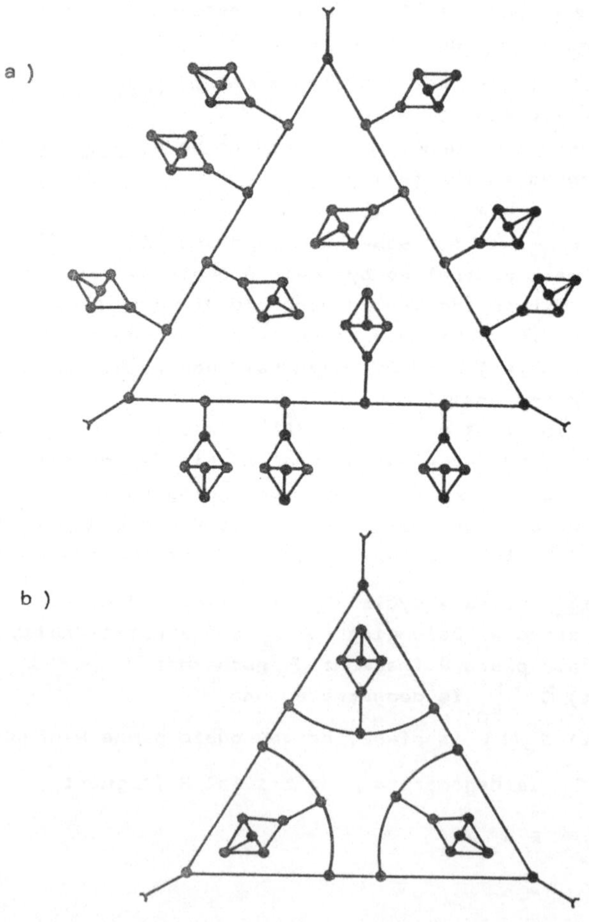

Fig. 24

If $\hat{w}$ starts and terminates with some $y_0 \in \{1,2\}$, one easily constructs a rotationally symmetric cubic R-fragment F satisfying ii) such that $\hat{w}$ fits F. This is shown in Fig. 24a for $\hat{w} = 121221$ or $\hat{w} = 211212$. The principle of this construction is obvious.

It follows that $\tilde{f}^{(F)}$ is degenerate for $f = \hat{w}^\infty$. For $\hat{w} = 12$, the fragment shown in Fig. 24b has this property.

<u>Claim.</u> Let $w_1, w_2 \in Y^*$, $w_1 \overset{k}{\longmapsto} w_2$ for $k \in \mathbb{N}$, and $f_1 = w_1^\infty$, $f_2 = w_2^\infty$. If $\tilde{f}_2^{(F)}$ is degenerate for some fragment F, then $\tilde{f}_1^{(F)}$ is degenerate, too.

This easily follows by induction on k.

The claim implies that $\tilde{f'}^{(F)}$ is degenerated for $f' = w'^\infty$ and the F constructed above. Now we take
$$a_0 = \delta^*(3^{\,\text{len}(w_0')}, a).$$
Then $\tilde{f}^{(F)}{}_{a_0} = \tilde{f'}^{(F)}$, i.e., the cycle $\tilde{c}^{(F)}{}_{a_0}$ is degenerate. For a better understanding, we remark that the state a does not necessarily belong to the cycle $\tilde{c}^{(F)}{}_{a_0}$. //

If we are considering more than one finite automaton and the state sets are not necessarily disjoint, the cycles of the state a with respect to the automaton α and $\tilde{\alpha}^{(F)}$ are denoted by $C_{\alpha,a}$ and $\tilde{C}^{(F)}{}_{\alpha,a}$, respectively.

<u>Proposition 1.</u> Let $\mathcal{A}$ be a finite set of finite R-automata. There is a rotationally symmetric cubic plane R-fragment F with three free half-edges such that

i) $\tilde{C}^{(F)}{}_{\alpha,a_0}$ is degenerate for each automaton $\alpha \in \mathcal{A}$ and any state a_0 of α, and

ii) $\sigma_F(L)$ is plane for any cubic plane R-graphoid L.

This is proved by an inductive procedure starting with

$$F_0 = \quad .$$

Let F_1 be defined for some $l \in \mathbb{N}$.

If no automaton $\alpha \in \mathcal{A}$ has a state a_1 such that $\tilde{C}^{(F_1)}{}_{\alpha,a_1}$ is non-degenerate, we stop the procedure with $F = F_1$.

Otherwise, let $\alpha_1 \in \mathcal{A}$ and a_1 be a state of α_1 such that $\tilde{C}^{(F_1)}{}_{\alpha_1,a_1}$ is non-degenerate. By Lemma 13, there are a state a_0 in $\tilde{C}^{(F_1)}{}_{\alpha_1,a_1}$ and an R-fragment $\bar{F}_1$ satisfying ii), and the cycle $\widetilde{c}^{(\overbrace{\widetilde{F_1}})^{(F_1)}}_{\alpha_1,a_0} = \widetilde{c}^{(\sigma_{F_1}(\overline{F}_1))}_{\alpha_1,a_0}$ is degenerate.

We define

$$F_{1+1} = \sigma_{F_1}(\overline{F}_1).$$

If $\widetilde{c}^{(F_1)}_{\alpha,a}$ is degenerate for some state a of an automaton $\alpha \in \mathcal{A}$, then $\widetilde{c}^{(F_{1+1})}_{\alpha,a}$ is also degenerate. Therefore, the number of degenerate (factorized) cycles of automata from $\mathcal{A}$ has increased by going from F_1 to F_{1+1}. Since the maximal number of cycles is bounded by the sum of the numbers of states of automata from $\mathcal{A}$, the procedure must stop in some step, and the obtained fragment F satisfies the assertions of Proposition 1. //

Now it is not hard to show that no 2-automaton system is able to search every cubic plane R-ficograph. For this purpose, let $\gamma = (\alpha_1, \alpha_2)$ be a system of two cooperating R-automata (of degree bound three), cf. Section 1.5 . By restricting the input alphabet of α_i, we obtain a finite R-automaton α_i' with the same behaviour as α_i if it is working alone ($i = 1,2$).

Let $F = (L_F, h_1, h_2, h_3)$ be a rotationally symmetric cubic R-fragment according to Proposition 1 for $\mathcal{A} = \{\alpha_1', \alpha_2'\}$. We consider the behaviour of the system γ starting on an image of h_1 in a hypervertex of position h_0 in $\sigma_F(\Gamma)$, where Γ is the embedded R-incograph shown in Fig. 4 a.

If the automata of γ see each other infinitely often, a trap for γ can be constructed like for a single automaton in the proof of Theorem 2, but using $\sigma_F(\Gamma)$ instead of Γ.

If α_1 and α_2 see each other only finitely often in $\sigma_F(\Gamma)$, then after a certain number of steps, each automaton α_i is working alone, i.e. according to α_i'. It finally reaches a cyclic state a_i and, since $\widetilde{c}^{(F)}_{\alpha_i,a_i}$ is degenerate, the automaton becomes cyclic on a finite part $\widetilde{L}_i$ of $\sigma_F(\Gamma)$. Taking from $\sigma_F(\Gamma)$ a sufficiently large finite part containing $\widetilde{L}_1$ and $\widetilde{L}_2$ and some vertex not visited by α_1 and α_2, we obtain a trap for the system γ.

The same argument obviously also applies to 2-head automata and even to jumping 2-head automata.

Moreover, the trap construction can analogously be transferred to k-automaton systems and k-head automata working in such a way that all automata or heads behave as they would be alone as long as they are not all on the same position. In other words, messages

are interchanged only at plenary meetings of all automata and
heads,respectively. Such systems or k-head automata are called
plenarily working,or briefly plenary.

More precisely,a k-head automaton
$$\alpha = (\overline{X} , \overline{Y} , A , \delta , \lambda , a_0) ,$$
as specified in Section 1.5 , is said to be _plenary_ if from
$P,P' \in \mathcal{P}_k$ and $P,P' \neq \{ \{1,2, \ldots ,k\} \}$ it follows that
$$\delta ((x_1, \ldots ,x_k,P) , a) = \delta ((x_1, \ldots ,x_k,P') , a)$$
and
$$\lambda ((x_1, \ldots ,x_k,P) , a) = \lambda ((x_1, \ldots ,x_k,P') , a)$$
for all $(x_1, \ldots ,x_k) \in X^k$ and all $a \in A$.

Any 2-head automaton is plenary,since $P,P' \in \mathcal{P}_2$ and
$\emptyset \neq P,P' \neq \{ \{1,2\} \}$ implies that $P = P' = \{ \{1\}, \{2\} \}$. Remark that
the behaviour of a plenary k-head automaton on regular R-graphoids
of some degree can be simulated by a plenary system of k cooperating automata.

The trap construction sketched above for 2-automaton systems
can immediately be transferred to plenary multihead automata.

Theorem 9. To any plenary multihead R-automaton,there is a trap
    (L,h) ,where L is a plane cubic R-ficograph. //

Obviously,one can construct a strong trap (L,h,v) with a plane
R-ficograph of degree bound three,and deg(v) = 1.

Moreover,there is a straightforward way,for a given vertex
substitution δ on R-ficographs,to define a k-head automaton
$\alpha(\delta)$ which,working on a labyrinth L, simulates the behaviour of
a k-head automaton α working on $\delta(L)$. The starting position of
this simulation can be arbitrarily fixed within the hypervertex
of the starting position of $\alpha(\delta)$. This can be defined analogously to the case of finite automaton , cf. Section 1.8 .

Therefore,the proofs of Theorem 7 and Corollary 9 can be modified to construct universal traps for sets of plenary multihead
automata.

Corollary 11. To any finite set of plenary multihead automata ,
    there is a universal strong trap (L,h,v) with a plane R-fico-
    graph L.
    To any infinite set of such automata , there is a universal
    mastering trap (L,h),where L is a plane R-incograph. //

<u>Hints & Sources.</u> Proposition 1 and the trap construction for two
cooperating automata are due to M. Blum and D. Kozen /L.BlKo/ .
Moreover,they sketched a trap construction for three cooperating
automata. In /L.Koz/ D. Kozen showed how this result can be improved to 4-automaton systems. These constructions make nontrivial
use of concepts of differential geometry. Unfortunately,a further
generalization of these methods to k-automaton systems for $k > 4$
seems to be hardly manageable. The corresponding problem is still
open.

3.7. Barrages and traps for cooperating systems

For the inductive construction of system traps,we need a
stronger version of the trap property. This leads to a more
powerful hypothesis of the induction step.

Let $\gamma = (\alpha_1 , \alpha_2 , \ldots , \alpha_k)$ be a system of k cooperating
(R- or C-) automata,as specified in Section 1.5 .

By a <u>barrage</u> for γ ,we mean a quadruple

$$B = (L , h_1 , h_2 , e)$$

consisting of an (R- or C-) ficographoid L with just two free
half-edges h_1 and h_2 and a bridge e of L such that h_1 and
h_2 belong to different components after cutting e ,and for any
(R- or C-) corridor $(\overline{L} , \overline{h}_1 , \overline{h}_2)$ disjoint to L the following
holds.

Let $\widetilde{L} = L \overset{+}{\vee}_{(h_1 ,\overline{h}_1 ; h_2 ,\overline{h}_2)} \overline{L}$; this means $\widetilde{L}$ is built by
joining L and $\overline{L}$ and gluing together the half-edge h_1 with
$\overline{h}_1$ and h_2 with $\overline{h}_2$.
If the automata of γ start with arbitrary states and on
arbitrary positions of $\overline{L}$ in $\widetilde{L}$, then no automaton reaches
the bridge e.
More precisely,for any two γ-configurations
$\varkappa' = (h_1' , a_1' , \ldots , h_k' , a_k')$ and

$$\varkappa'' = (h_1'' , a_1'' , \ldots , h_k'' , a_k''),$$

where $h_i' \in H_{\overline{L}}$ for all $i \in \{1,\ldots,k\}$ and $\varkappa''$ is obtained
from $\varkappa'$ by applying the configuration transition of
a certain number of times , it holds $h_i'' \notin e$ for all
$i \in \{1, \ldots , k \}$.

The construction of a barrage B and a labyrinth
$\tilde{L} = L \overset{+}{\vee}_{(h_1, \overline{h}_1 ; h_2, \overline{h}_2)} L$ is illustrated in Fig. 25 .

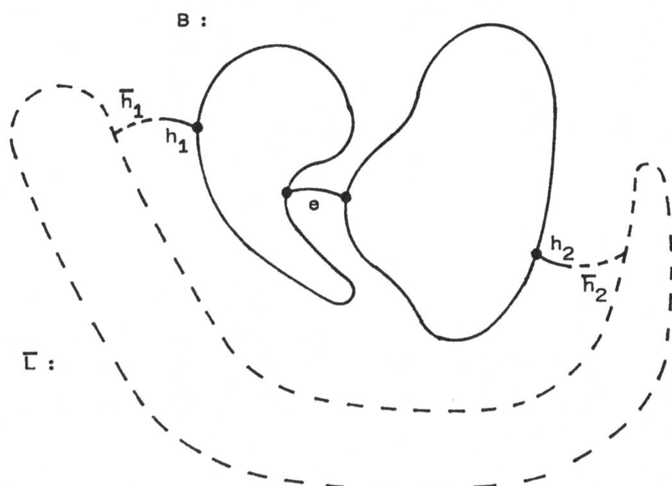

Fig. 25

The barrage B is said to be <u>planar</u> if the graphoid underlying
L has a 2-dimensional embedding such that both ang(h_1) and
ang(h_2) belong to exterior faces. It is called <u>plane</u> if there is
such an embedding which corresponds to the rotation system of L.

By a <u>universal</u> barrage for some set $\mathcal{S}$ of cooperating systems,
we mean a quadruple B being a barrage for each system σ from $\mathcal{S}$.

<u>Theorem 10.</u> To any finite set of systems of cooperating R-automa-
ta , there is a universal planar barrage.

We inductively construct so-called <u>(k,m)-barrages</u> . These
are universal barrages for the set of all cooperating systems of
at most k automata,each with at most m internal states. Let the
degree bound three be fixed.

Planar (1,m)-barrages can be obtained from Theorem 7. Indeed,
the set of 1-automaton systems with $\leq m$ states is equivalent to
a finite set of finite R-automata. By Theorem 7 , there is a
universal strong trap (L' , h' , v') for all single finite automata with $\leq$ m+1 internal states. We have deg(v') = 1 , and the
graph underlying L' possesses a 2-dimensional embedding P whose
exterior region touches both P(h') and P(v').

After cutting the edge $\{h', hal(h')\}$ in L' , let L_1' denote
the component containing v' , and let the half-edge h_1' from

{h' ,hal(h')} belong to L_1'. Finally,we construct L by joining
two disjoint isomorphic copies of L_1' via an edge e connecting
the images of vertex v'. The two free half-edges (these are the
images of h_1') are denoted by h_1 and h_2 , cf. Fig. 26 .

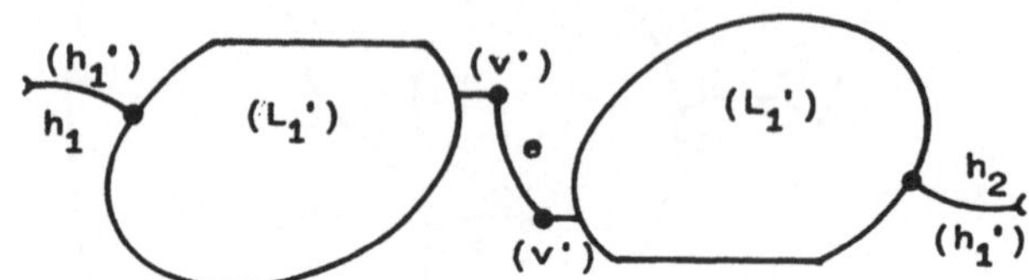

Fig. 26

<u>Claim.</u>   (L , h_1 , h_2 , e) is a planar (1,m) - barrage.

The planarity is obvious. Assume that , for some disjoint cor-
ridor $(\Gamma,\overline{h}_1,\overline{h}_2)$, a single R-automaton α with m states would
reach the bridge e if it starts on a half-edge from Γ in the
labyrinth $\tilde{L} = L \overset{\text{v}}{\underset{(h_1,\overline{h}_1;h_2,\overline{h}_2)}{}} \Gamma$.

By t_1 we denote the first point of time in which α reaches a
position in the edge e. There is a last moment $t_0 < t_1$ in which
α occupies a position h_i for $i \in \{1,2\}$. Within the time interval $[t_0, t_1]$,the automaton is working in (a copy of) L_1' , and
it traverses this graphoid from h_1' to vertex v'.

Let a_0' be the state of α at the time t_0. If $h_1 = h'$, we define
$\alpha' = \alpha$. If $h_1 = hal(h')$,let α' be the automaton with m+1 states
which at the start (on h') moves backward (to h_1) and enters the
state a_0',and continues its work according to α then.

Obviously,starting on the half-edge h' in the labyrinth L' ,
α' would reach the vertex v'. This contradicts the construction
of (L',h',v') according to Theorem 7.

The claim has been proved.

For the induction step , let $k \geqslant 1$ and

$$B_k = (L_k , h_{k1} , h_{k2} , e_k)$$

be a planar (k,m)-barrage. By F' we denote the 2k th power of the
corridor (L_k , h_{k1} , h_{k2}),

$$F' = (L_k , h_{k1} , h_{k2})^{2k} ,$$

and σ is the edge substitution inserting F' into all edges of
R-graphoids ,

$$\sigma = \{ e \longrightarrow F' \} .$$

Now we take a planar $(1,m^\times)$-barrage $B^\times = (L^\times , h_1^\times , h_2^\times , e^\times)$
for a sufficiently large $m^\times$ which will be specified later on .

186

We shall show that
$$B = (\,\sigma(L^*)\,,\,h_1^*\,,\,h_2^*\,,\,\hat{e}\,)\,,$$
where $\hat{e}$ denotes the middle edge within the hyperedge inserted into e^*, is a planar $(k+1,m)$-barrage. B is sketched in Fig. 27, where L_1^* and L_2^* denote the two parts from which B^* is built, cf. Fig. 26. By v_1 and v_2, we mean the vertices connected by e^* in L^*.

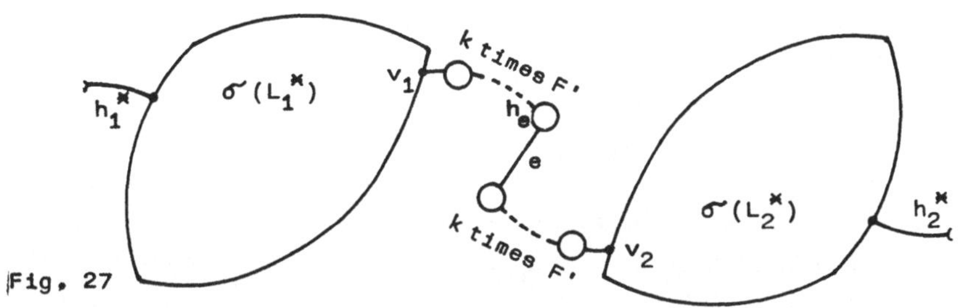

Fig. 27

The planarity of B is obvious from our construction.

To prove the barrage condition, let
$$\gamma = (\,\alpha_1\,,\,\ldots\,,\,\alpha_{k+1}\,)$$
be a system of k+1 cooperating R-automata, each with at most m states. Assume that there is an environmental corridor $(\Gamma,\overline{h}_1,\overline{h}_2)$ such that an automaton from γ reaches the edge e in the labyrinth
$$\tilde{L} = L \overset{+}{\underset{(h_1^*,\overline{h}_1\,;\,h_2^*,\overline{h}_2)}{\vee}} \Gamma$$
if all automata start in certain states on positions from Γ.

Let t_1 denote the first point of time at which an automaton reaches e.

For $0 \leqslant t \leqslant t_1$, P_t denotes the set of positions of the automata of γ at the time t in L. We say that the system γ is <u>separated</u> at the time t if there is a decomposition of P_t into two non-empty subsets P_{t1} and P_{t2} (i.e. $P_t = P_{t1} \cup P_{t2}$,

$P_{t1} \cap P_{t2} = \emptyset$, $P_{t1} \neq \emptyset \neq P_{t2}$) such that any path connecting a position from P_{t1} with a position from P_{t2} must traverse a copy of L_k. The latter means that any such a path must contain the images of h_{k1} and h_{k2} in some copy of L_k contained in $\tilde{L}$.

In this case, the decomposition $P_t = P_{t1} \cup P_{t2}$ defines a decomposition of the automata from γ into two sets γ_{t1} and γ_{t2}, where γ_{t1} consists of all automata with positions from P_{ti}

at the time t (i = 1,2).

<u>Lemma 14.</u> Let the system γ be separated at some time $t \in [0, t_1]$,
where $\mathcal{S}_{t\,1}$ and $\mathcal{S}_{t\,2}$ be the corresponding sets of automata.
Then there is no point of time $t' \geqslant t$ at which an automaton
from $\mathcal{S}_{t\,1}$ has the same position as an automaton from $\mathcal{S}_{t\,2}$;
i.e., no automaton from $\mathcal{S}_{t\,1}$ can see an automaton from $\mathcal{S}_{t\,2}$,
and conversely.

Indeed, within the time interval $[t, t_1]$ the automata from
$\mathcal{S}_{t\,1}$ and $\mathcal{S}_{t\,2}$, respectively, are working separately as long as they
do not see each other. Thus their working can be thought to be
performed by suitable systems γ_1 and γ_2 consisting of
$k_1 = \mathrm{card}(\mathcal{S}_{t\,1})$ and $k_2 = \mathrm{card}(\mathcal{S}_{t\,2})$ cooperating automata,
respectively. Since $k_1, k_2 \leqslant k$, none of these systems can reach an
image of the bridge e_k in an L_k-copy which does not contain an
automaton at the time t. Therefore, no automaton from $\mathcal{S}_{t\,1}$ can
meet one from $\mathcal{S}_{t\,2}$. //

From the lemma it follows that if the system γ is separated
at some time $t \in [0, t_1]$, then after that moment no automaton can
traverse an L_k-copy which does not contain an automaton at the
time t.

Without loss of generality, we assume that the system γ reaches
the edge e via $\sigma(L_1^{\times})$. Let t_2 be the first point of time in
which an automaton occupies the position h_e, that is the entrance
of the F'-copy whose exit belongs to e, see Fig. 27. Since γ
reaches e, it cannot be separated at a time $t < t_2$.

Let t_0 be the last point of time $\leqslant t_2$, at which an automaton
of γ, say α_1, occupies the position $h_1^{\times}$. During the time interval $[t_0, t_2]$ the system γ is working within $\sigma(L^{\times})$.

Moreover, it can be simulated by a single R-automaton α_γ
working on $L^{\times}$. For this purpose, to an γ-configuration
$$\varkappa = (h_1, a_1, \ldots, h_{k+1}, a_{k+1}),$$
we assign the α_γ-configuration
$$(\hat{h}_1, \hat{a}_\varkappa),$$
where $\hat{h}_1$ be the $L^{\times}$-position that is nearest to h_1' in $\sigma(L^{\times})$,
and $\hat{a}_\varkappa$ describes the configuration $\varkappa$ with respect to $\hat{h}_1$ in
some fixed way. Without going into details, we remark that
$$(3 \cdot 2k \cdot \mathrm{card}(H_{L_k}) \cdot \mathrm{card}(A))^k$$
internal states of α_γ are sufficient for this simulation if A
denotes the set of all states of automata of γ.

Since γ reaches e via $\sigma(L_1^*)$, α_γ reaches the vertex v_1 finally if it starts with a state coding the γ-configuration at the time t_0. Therefore, we could construct a finite R-automaton α^* with

$$m^* = 4\,(\,6\,k\cdot\mathrm{card}\,(\,H_{L_k}\,)\cdot\mathrm{card}\,(\,A\,)\,)^k + 1$$

internal states such that it reaches the bridge e^* in L^* if it starts on position $\mathrm{hal}(\,h_1^*\,)$ in the labyrinth

$$L^* \overset{+}{\underset{(h_1^*,\,\overline{h}_1^*\,;\,h_2^*,\,\overline{h}_2^*\,)}{v}} L^*$$

for an arbitrary corridor $(\,L^*,\,\overline{h}_1^*,\,\overline{h}_2^*\,)$ disjoint to L^*. Indeed, starting on $\mathrm{hal}(\,h_1^*\,)$, let α^* go back to position h_1^* and simulate the automaton α_γ then, but at any reached vertex, let it additionally visit all (three) neighbouring vertices.

The existence of α^* contradicts the supposition that B^* is an $(\,1,m^*\,)$-barrage if m^* is defined as given above.

This completes the proof of Theorem 10. //

<u>Corollary 12.</u> To any finite set of systems of cooperating
    R-automata, there is a universal strong trap $(\,L,h,v\,)$,
    where L is a planar R-ficograph of degree bound three,
    and $\deg(\,v\,) = 1$.
    To any set of cooperating systems, there is a universal
    mastering trap (L,h) with a planar R-incograph L of degree
    bound three.

Indeed, from a barrage $(\,L,h_1,h_2,e\,)$ one obtains a strong trap (L,h,v) by inserting the corridor ⊢•⊣ into the edge e and connecting h_1 with h_2 by the corridor .

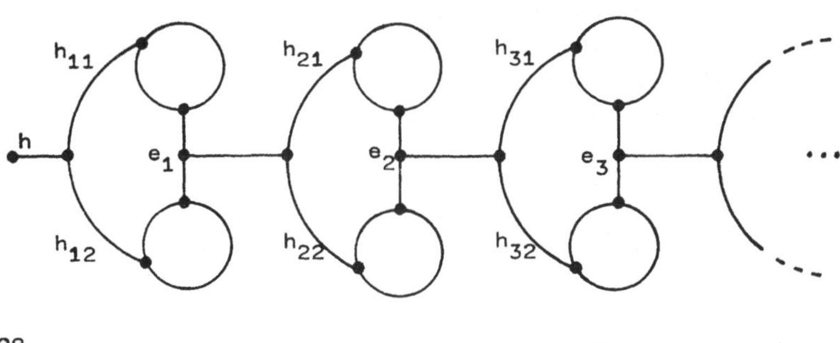

Fig. 28 B_1 B_2 B_3 ...

A planar universal trap for the set of all cooperating systems
can be constructed as shown in Fig. 28 , where
$$B_k = (L_k , h_{k1} , h_{k2} , e_k)$$
denotes a planar (k,k)-barrage for $k \in \mathbf{N}^+$. //

The reader may already have remarked that a single face-following automaton does not possess a plane or a 2D barrage. The
general problem of the existence of finite plane traps for cooperating R-automata is still open , as mentioned in the previous
section , whereas 2D ficographs can be searched by two cooperating
C-automata , cf. Theorem 2.9 .

By a (strongly normed) <u>3D barrage</u> for some system γ of cooperating finite C-automata whose direction set contains $\mathbf{D}_3$,
we mean a barrage
$$B = (L , h_1 , h_2 , e)$$
for γ ,where L is a C-graphoid having a (strongly normed)
3D embedding P such that the set

    $P (h_1) \cup P (h_2)$ belongs to the same straight line , and

    $P (h_i) \not\subseteq \text{box} (P(V_L))$ for $i = 1,2$.

This means,the "pins" h_1 and h_2 lie on a straight line,but their
free ends are outside the box generated by the (embeddings of
the) vertices of L .

<u>Theorem 11.</u> To any finite set of cooperating C-automata whose
    direction set contains $\mathbf{D}_3$, there is a universal strongly
    normed 3D barrage.

The proof of this theorem is analogous to that of Theorem 10.
It is completely given in /L.He87a/ (with some slight modifications of the notations). So we only sketch how strongly normed
3D (1,m)-barrages can be obtained.

For this purpose , let (L,h,v) be a universal strong trap for
the set of all C-automata with a direction set containing $\mathbf{D}_3$ and
at most m+1 internal states. Moreover,let (L,h,v) satisfy the
conditions given in Theorem 8 , for j = 2 .

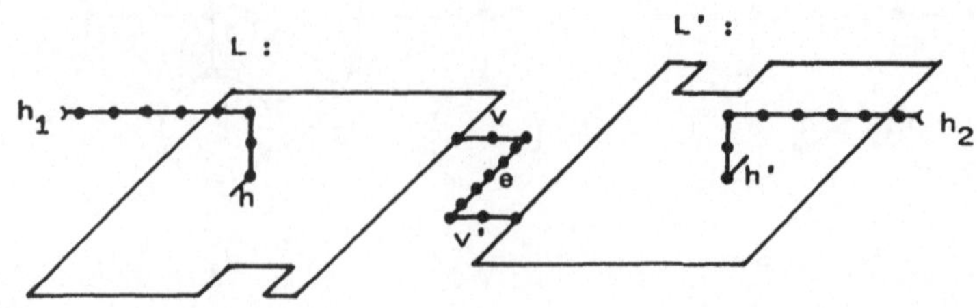

Fig. 29

190

Fig. 29 shows how a strongly normed 3D $(1,m)$-barrage can be
constructed by suitably joining L with a copy L' of itself
rotated by an angle of 180°. Of course, such a rotation preserves the universal-trap property. //

Similarly to Corollary 12, it follows

<u>Corollary 13.</u> To any finite set of systems of cooperating
C-automata whose direction sets contain D_3, there is a universal strong trap (L,h,v), where L is a strongly normed
3D ficograph.

To any set of such systems, there is a universal mastering
trap (L,h) with a strongly normed 3D incograph L.

Indeed, first one obtains universal normed 3D traps by a proof
analogous to that of Corollary 12. These are made strongly normed
using a simple edge substitution analogously to that in the proof
of Lemma 1.13 . //

Finally, we remark that the concept of barrage makes no sense
for multihead automata. Let be given a system barrage

$$B = (L, h_1, h_2, e).$$

There is always an environmental corridor $(\bar{L}, \bar{h}_1, \bar{h}_2)$ which
represents, in some suitably coded form, a control word for a
path from h_1 to the bridge e. Then, starting on position $\bar{h}_1$ in
the labyrinth

$$\tilde{L} = L \overset{\dot{v}}{}_{(h_1, \bar{h}_1 \; ; \; h_2, \bar{h}_2)} \bar{L} \; ,$$

one head of a 2-head automaton can read this code and control the
second head from h_1 to e.

<u>Hints & Sources.</u> The main idea of this section, namely the in-
ductive construction of (k,m)-barrages, is due to M. Rabin (unpublished). Theorem 10 was obtained by H.-A. Rollik /L.Ro/ .
Theorem 11 was claimed by M. Blum and W. Sakoda /L.BlSa/. They
also sketched a proof. A complete proof and some modifications
of this theorem were given in /L.He87a/ , see also /L.He87c/.

CHAPTER IV. SUPPLEMENTS AND PROBLEMS

The aim of this fragmentary last chapter is to complete our
survey of labyrinth research by reporting some further results
both on searching problems and on other questions connected with
labyrinths and by stating some open problems which seem to be
important for progress in research.

4.1. Automata without markers in finite labyrinths

By Corollary 3.1 , to every finite automaton with m states,
there is a plane R-trap with two faces and $O(m)$ vertices. This
trap can be assumed to be quasi-regular of degree 3 , and analogously to the proof of Corollary 3.4, the result can be generalized
to finite C-automata whose direction set contains at least three
elements. We obtain plane C-traps with two faces and $O(m)$ vertices.

For finite C-automata with a direction set containing D_2, by
Corollary 3.7 , we have strongly normed 2D traps with only three
faces but exponentially many vertices. Some questions remain
open.

<u>Problem 1.</u> Does there exist a finite C-automaton (or 1-counter
    or 1-pushdown automaton) which searches all 2D ficographs
    with only two faces ?

For finite automata, one can equivalently consider the
(strongly) normed 2D ficographs instead of the 2D ficographs.
In this form, the 2-component problem is due to G. Asser /L.As/

and H. Müller /L.Mu79/. By **Theorem** 2.7 , we know that there is a
1-counter automaton searching all normed 2D ficographs and halting finally.

   The exponential size of Budach's trap already results from the
construction in the proof of Proposition 1.3 . As shown in
Lemma 3.11 , any size estimation of traps for finite C-automata
yields a lower bound of the space complexity of the searching of
(strongly normed) 2D ficographs. Especially,a polynomial size of
the traps would yield the lower bound log(n). Of course,such a
trap construction would require a technique which essentially
differs from Budach's method.

<u>Problem 2.</u> Does there exist a polynomial p such that,to any
     finite C-automaton with m states and a direction set con-
     taining $\mathbb{D}_2$, there is a 2D trap with only p(m) vertices ?

   Independently on this approach , we would like to stress the
problem of improving the upper and/or lower bounds of the space
complexity of searching the finite d-dimensional mazes. By
Theorem 3.6 , we have the lower bound log ∘ log(n). From Theorems
2.11 and 2.12 , one obtains the upper bounds log(n) , for d = 2 ,
and n , for d ⩾ 3 .

<u>Problem 3.</u> Improve the lower and/or upper bounds of the space
     complexity of searching all finite d-dimensional **mazes**.

   For halting automata , we can state a better lower bound.

<u>Supplement 1.</u> If the strongly normed 2D ficographs (with two
     faces) can be searched by an s(n) space-bounded Turing tape
     automaton which always halts after the search,then
          log(n) = O (s(n)) .

   The proof is essentially given in /L.He83 a/ .

   By Corollaries 3.2 and 3.3 and Theorem 3.2 , already rather
restricted types of R-ficographs cannot be searched by 1-pushdown
automata or space-bounded Turing tape automata. This also holds
for the plane cubic C-ficographs with a direction set of at least
three elements. The corresponding question for (non-normed) 2D or
3D ficographs is open.

<u>Problem 4.</u> Can the 2D or 3D ficographs be searched by a space-
     bounded Turing tape automaton ?

   By the proof of Theorem 3.3 , such an automaton cannot halt
after the search.

   It is possible to modify the definition of searching a labyrinth by a Turing tape automaton in such a way that the automaton

starts with an information on its worktape about the size of the
labyrinth in which it is put. Then the R-ficograph (of some
degree bound) can be searched by a linearly space-bounded automaton which always halts finally. For details the reader is referred
to /L.He82/ .

In this context,we also have to mention the problem of short
traversal sequences which is closely related to the space complexity of labyrinth searching. It was put by S. Cook in connection with the reachability problem for undirected graphs. This is
the analogue to the reachability problem for directed graphs which
is logspace-complete for the complexity class NSPACE (log(n)) ,
cf. /L.AlKaLiLoRa/ , /S.Sav70,73/ .

A control word $w \in \{0,1,...,b-1\}^*$ is said to be an <u>L-universal
traversal sequence</u> , for some R-ficograph L , if the walk controlled
by w on L visits all vertices of L, for any starting position.
w is called <u>n-universal</u> , for $n \in \mathbb{N}^+$, if it is L-universal for any
R-ficograph L having just n vertices. These definitions can be
straightforwardly transferred to (regular) C-ficographs of some
fixed direction set.

One easily obtains L-universal traversal sequences of length
$O(n_L^3)$, see /L.He82/. In /L.AlKaLiLoRa/ , n-universal traversal
sequences of length $O(n^3 \cdot log(n))$ were obtained by means of
probabilistic methods. In /L.KaPaSi/ , traversal sequences universal for the class of all n-vertex cliques are constructed with
the length $n^{O(log(n))}$.

<u>Problem 5.</u> Construct short universal traversal sequences for our
types of labyrinths,or give non-linear lower bounds of the
length of universal traversal sequences.

By Theorem 2.7 , there is a 1-counter automaton which searches
all normed 2D ficographs L and halts finally. The proof shows
that the counter content is even bounded by n_L. Whereas the plane
cubic C- or R-ficographs cannot be searched by a 1-counter automaton , the question is still open for 2D ficographs.

<u>Problem 6.</u> Does there exist a 1-counter automaton which searches
all 2D ficographs ?

This is connected with Problem 1. Note also Theorem 2.8 in
this context.

Now we consider universal traps for finite sets of finite automata. As mentioned after the proof of Theorem 3.7 , the universal
(strong) traps for finite sets of R- or C-automata,as they are
constructed in the proofs of Theorems 3.7 and 3.8 , have

exponentially many vertices and faces with respect to the cardinalities of the given sets. Also Hoffmann's /L.Ho81,82/ construction technique for universal traps with parallel tests does not
yield better results.

Problem 7. Can universal (strong) traps be constructed with
polynomially many vertices and/or faces or even with a
face number bounded by a constant , for all finite sets of
finite automata ?

For C-automata , there are connections to Problem 2.
Another question concerns the size of the faces , i.e. the
numbers of angles in them. By Theorem 2.10 , the cardinalities of
the faces in 2D traps cannot be universally bounded , but for
R-automata this is possible , as Theorem 3.2 shows.

Problem 8. Can the size (i.e. the cardinality) of the faces
of universal (strong) traps for finite sets of finite
R-automata be bounded by some constant ?

Finally,we consider searching problems for halting automata.
The finite labyrinths with only one face , these are the trees ,
can be searched by a finite automaton which never halts , see
Theorem 2.2 ; but they cannot be searched by a halting finite automaton , see Theorem 3.4 and Corollary 3.4 . As already mentioned
in Section 3.2 , one easily sees

Supplement 2. There is a 1-pushdown automaton which searches
every finite R-tree (of some fixed degree bound) and
always halts finally.

This is shown in /L.BulHe/. The corresponding question for
1-counter automata is still open.

Problem 9. Does there exist a 1-counter automaton which searches
all finite R-trees or 2D trees and always halts finally ?

For an additional remark concerning centred trees , see the end
of Section 3.2 . Problem 9 and Problem 1 are connected with a
question asked by M. Blum and D. Kozen /L.BlKo/ , namely

Problem 10. Does there exist a 1-counter automaton which , if it
is put in a normed 2D ficograph , searches the face to the
left of its starting position and halts on its starting
position finally ?

Analogously to Theorem 3.3 , one shows that this is impossible
on all 2D ficographs (with two faces). In /L.BuMi/ , L. Budach
and U. Mieth gave a negative answer to Problem 10 for 1-counter

automata which never move back (i.e. for forward automata) ;
the general problem is still open .

4.2. Marker automata and cooperating systems in finite labyrinths

By Program 2.1 and Theorem 2.1 , we know that the R-ficographs
L (of some fixed degree bound) can be searched by a pointer
automaton using $O(n_L)$ indistinguishable pointers and halting
after $O(n_L)$ steps. We have shown a related result for pebble
automata.

<u>Supplement 3.</u> There is a pebble automaton which searches any
    R-ficograph L (of some fixed degree bound), halts after
    $O(n_L^2)$ steps and uses $O(n_L)$ indistinguishable pebbles.

The proof is still unpublished. Using P. Schreiber's variant
of C. Wiener's searching algorithm , see /L.Sc/ , /L.Wi/ , /L.Ko/ ,
one obtains an algorithm which can be implemented by a pebble
automaton in exponential time. By some additional effort , the
polynomial time complexity can be secured. It is unknown whether
linear time complexity can be reached. This is possible on the
2D ficographs by means of a modification of Tarry's algorithm.

Theorem 2.3 gives an example how the number of used markers
(pointers) can be bounded by a certain non-trivial complexity
measure of the labyrinth L, namely the number of faces. Then the
time complexity increases to $O(n_L^2)$. It is not known whether
linear time complexity is possible here , or whether a related
result for pebble automata holds. We summarize these questions **by**

<u>Problem 11.</u> Does there exist a pebble automaton searching all
    R-ficographs , halting within linear time and using only
    O(f) pebbles on R-graphs with f faces ?

For the degree bound three , we believe to be able to give an
$O(n_L^2)$ - time algorithm using only O(f) pebbles (unpublished).

A problem which should be mentioned here is that of cleaning
up the labyrinth after the search. By this we mean that the automaton should carry all its markers when it finishes its work.
In a certain contrast to this , H. Müller /L.Mu77a/ constructed
a non-erasing searching automaton. This cannot remove a marker
from a place on which it has been put some time.

196

Now we consider k-marker automata and cooperating systems. The
strongest generalization of Budach's trap-construction technique
has been given by F. Hoffmann who has stated

Supplement 4. There is no 1-pebble automaton searching all
    strongly normed 2D ficographs.

Remember that one pebble is equivalent to one pointer on 2D
graphs. Hoffmann's proof is based on a sophisticated discussion
of the behaviour of a 1-pebble automaton in certain test labyrinths constructed by means of Budach's technique and suitable
interlinking operations. Unfortunately, in /L.Ho81/ only a sketch
of the proof is given, and also in the presentation in /L.Ho82/
many details are left to the reader. So we neither could present
a proof of this interesting result within this book, nor were we
able to verify completely the claim.

Problem 12. Give a complete and possibly shorter proof of
    Supplement 4.

It is relatively simple to show that also for 1-pebble automata the problem of searching all strongly normed 2D ficographs is
equivalent to the problem of searching all 2D ficographs. Moreover, from Supplement 4 one obtains universal (strong) traps for
sets of 1-pebble automata by means of vertex substitutions quite
analogously to Theorem 3.8 and Corollary 3.10 .

In contrast to Supplement 4 , by Theorems 2.7 and 2.9 , the
normed 2D ficographs and even all 2D ficographs can be searched
by a 2-pebble automaton or by two cooperating automata. Theorem
3.9 says that two cooperating automata cannot search the plane
cubic R-ficographs. This can easily be generalized to plane cubic
C-ficographs.

Theorem 2.9 also deals with (1-pebble,1-counter) automata. We
don't know any technique for obtaining traps for such mixed types
of automata.

Problem 13. Does there exist a (1-pointer,1-counter) automaton
    which searches any plane cubic R-ficograph ?

Another complementary remark to Supplement 4 follows from
Corollary 2.4 and Theorem 2.6 . They show that the numbers of
faces or of vertices of orders $\geqslant 3$ in traps for 1-pointer automata
cannot be bounded by a constant.

The number of internal states of the 1-pointer automaton from
Theorem 2.6 exponentially depends on the number k of vertices
of orders $\geqslant 3$.

Problem 14. Can the first part of Theorem 2.6 be proved by con-
    structing a 1-pointer automaton with only polynomially many
    states , with respect to k ?

Note that it is not known whether the first part of Theorem 2.6
can be proved by means of an always halting 1-pointer automaton.
Also the searching time needed by the automaton from Theorem
2.6 exponentially depends on the size of the given labyrinth.
Generally,the problem of reducing the time complexity is poorly
treated in Chapter II. On the other hand , there is a lack of techniques for proving non-trivial lower time bounds.

Problem 15. Improve the time complexities of the searching
    algorithms from Chapter II and/or give non-trivial lower
    bounds of the time complexities of the corresponding problems.

In /L.BlKo/ M. Blum and D. Kozen conjectured that there is no
finite system of cooperating finite automata which search all
cubic plane R-ficographs. So far it has only been shown that four
cooperating automata cannot do this task , see /L.Koz/. The general
question is open for marker and multihead automata,too.

Problem 16. Does , for some $k \in \mathbb{N}^{+}$, there exist a k-pointer au-
    tomaton or a k-automaton system or a (jumping) k-head au-
    tomaton searching all cubic plane R-ficographs ?

By Theorems 3.10 and 3.11 , we only know that the planar
R-ficographs and the (strongly normed) 3D ficographs cannot be
searched by a cooperating system. For multihead automata even
such results are not known.

Problem 17. Does there exist a multihead automaton searching all
    (strongly normed) 3D ficographs or all planar R-ficographs
    (of some fixed degree bound) ?

Only the analogue to Rabin's original result can be shown for
multihead automata.

Supplement 5. There is no jumping multihead automaton which
    searches all C-ficographs of some fixed direction set con-
    taining at least three elements.

This was proved by S. Cook and C. Rackoff /L.CoRa/. The proof
is surprisingly complicated. It is based on an estimation of the
numbers of visited vertices in certain "completely homogeneous"
graphs.
So it remains an interesting task to develop a technique for
constructing traps for multihead automata,possibly in a manner
like introduced by M. Rabin for cooperating systems.

We know that multihead automata are more powerful than cooperating systems , since they do not allow to construct barrages ,
cf. the end of Section 3.7 . With respect to searching problems,
this question is unsolved.

<u>Problem 18.</u> Are , with respect to searching problems , multihead
automata more powerful than finite cooperating systems , and
are these more powerful than finite automata with finitely
many markers ?

In this context , one should deal with the question about the
relationships between automata with distinguishable and such with
indistinguishable markers , respectively. There is a nice result
found by J. Ulehla.

<u>Supplement 6.</u> Let α be a k-pebble C-automaton with distinguish-
able pebbles , for the direction set D_2 .
Then there is a k-pebble automaton α' using indistinguishable pebbles such that the following holds. Starting with
all its pebbles on some position in an arbitrary normed 2D
graph L, α' simulates the behaviour of α starting with its
pebbles on the same position in L.
The analogous result holds for pebble automata in normed 3D
graphs , and it can be transferred to any higher dimension.

Unfortunately , no proof has been published so far , although
the result was announced more than five years ago and a proof
could be given on few pages. The basic idea is that any pebble i
of α $(1 \leq i \leq k)$, when it is deposited on a vertex v_i , is simulated
by a corresponding pebble i' of α'. In order to store the additional information about the colour of pebble i , the corresponding α'-pebble is deposited on a vertex v_i' connected with v_i
via a simple path w_i containing at most k vertices. The paths
$w_1, \ldots, w_k$ are stored in the state of α' ; moreover it keeps track
modulo k the differences between the (coordinates of the)
current position and the vertices v_i' $(1 \leq i \leq k)$. One can secure
that these differences do not coincide modulo k for different
pebbles.
The precise elaboration of this technique and the proof of
correctness are not hard if the idea is known. Remark that the
proof of Supplement 6 for the 2-dimensional case is even simpler.
The idea sketched above is due to K. Kriegel , and we are grateful
to him for the communication.
In /L.Ul/ , J. Ulehla showed that for two cooperating automata
using two pebbles in 1-dimensional labyrinths an analogous result

does not hold.

For k-marker automata on non-normed labyrinths,the corresponding question is open.

<u>Problem 19.</u> Can k indistinguishable markers (pointers,pebbles)
     simulate k distinguishable ones , or are the latter more
     powerful than the former , in non-normed C-graphs or in
     (plane) R-graphs ?

4.3. Automata in infinite labyrinths

On a first view , infinite labyrinths seem to be less important.
Indeed , the natural structures to which we could apply labyrinth
results are finite in most cases. But even if one is preferably
interested in finite objects,it could be useful to consider infinite , too. So we have seen in Section 1.6 that there are close
relationships between searching problems of finite labyrinths and
mastering problems for corresponding types of infinite labyrinths.

For multihead automata , the searching problem for finite labyrinths is even equivalent to the searching problem for infinite
ones .

<u>Supplement 7.</u> If there is a jumping k-head automaton searching
     all finite labyrinths of a certain kind,then there is a
     jumping (3k+2)-head automaton searching all infinite laby-
     rinths of the corresponding type.
     If there is a k-head automaton (without jumps) which searches
     all finite labyrinths of a certain type , then there is a
     (4k+1)-head automaton searching all infinite labyrinths of
     the corresponding type.

We will not give an explicit definition of what is meant by
"the corresponding type" ; the following simple idea of proof will
show what is essential.

For example , let α be a jumping k-head automaton searching
all cubic plane R-ficographs. If it starts with all its heads on
some position in a cubic plane R-incograph L , α would master
this labyrinth , i.e.,it would reach infinitely many configurations.
This fact can be applied to simulate a 2-counter automaton in a
similar way as described in the proof of Theorem 2.13 .

Indeed,let one head of a (3k+2)-head automaton α' be fixed

at the starting position v_0. Two blocks of k heads each are used
to represent the contents of two counters. More precisely , the
counter content z is represented if the k heads of the corresponding block occupy the positions of the heads of α reached
after z work steps if it starts on v_0 ; the internal state of α
in this configuration may be stored as a component of the state
of α'. The increase of the counter and the test for zero can be
easily implemented. To decrease z , one uses the third block of k
heads of α' which simulates the behaviour of α starting on v_0
once more. For this purpose,all these heads first have to jump
to v_0. So the α-configuration preceding that one representing z
can be computed.

Therefore , using 3k+1 heads , any 2-counter automaton can be
simulated. Now we can take a 2-counter automaton which generates
a control sequence causing the (3k+2)nd head to search the whole
labyrinth.

In this idea,we essentially used the jumping ability of heads,
namely in order to decrease the counter value. It is well-known
that a counter can be simulated by two non-decreasing counters
if the test of the equality of the two counter values is allowed.
Indeed,the value of the simulated counter can be represented by
the difference of the values of the two simulating counters. To
decrease this value,one has to increase the lower counter.

If α is a (non-jumping) k-head automaton which searches all
cubic plane R-ficographs , then on cubic plane R-incographs , a
non-decreasing counter can be represented by a block of k non-
jumping heads simulating α. A multihead automaton using some of
such blocks can recognize whether two blocks occupy the same positions. In this way , a (non-jumping) (4k+1)-head automaton can
simulate two counters by its four blocks , and it can cause the
remaining (4k+1)st head to search the given cubic plane R-incograph.

By Supplement 7 , the problem of whether there is a multihead
automaton searching all finite labyrinths of a certain type is
equivalent to the problem of searching all infinite labyrinths
of the corresponding type by a multihead automaton. Especially ,
this holds for cubic (plane) R-graphs and for the 2D or 3D graphs.
May be that solutions of Problems 16 and 17 become possible by
this observation.

It is not known whether analogous relationships hold for cooperating systems or marker automata.

<u>Problem 20.</u> Can the cubic plane R-incographs be searched by
            finitely many cooperating automata ? Is this problem equiv-
            alent to that of searching all cubic plane R-ficographs by
            a cooperating system ? What about the analogous problems
            for R-automata with finitely many (distinguishable or in-
            distinguishable) markers ?

By Theorem 2.13 , we know that the 2D incographs can be searched
by a 7-pebble automaton (with distinguishable pebbles) , by 7 cooperating automata or a 5-head automaton. Theorem 2.15 says that
there are a searching 5-pebble automaton and a 5-automaton system
for all normed 2D incographs. A strong lower bound is known only
for the normed case.

<u>Supplement 8.</u> There is no 4-pebble automaton and no system of
            four cooperating automata which search all strongly normed
            2D incographs.

This has been stated by F. Hoffmann , and an idea of proof is
sketched in /L.Ho84a,b/. This proof uses the same technique as
that of Supplement 4 and is even based on that result. So the
comment given there applies here , too.
    F. Hoffmann has shown that the co-finite 2-dimensional mazes ,
or the corresponding incographs , can be searched by four cooperating automata , and he puts the following question.

<u>Problem 21.</u> Can the co-finite 2-dimensional mazes be searched by
            three cooperating automata ?

Concerning 2D incographs we propose

<u>Problem 22.</u> Improve the upper bounds given above for searching
            all 2D incographs and/or give non-trivial lower bounds.

Especially the multihead automata have not yet been considered
with respect to these questions ; also Problem 21 could be asked
for 3- or even 4-head automata .
The problem of searching the whole discrete plane , i.e. the 2D
incograph with the vertex set Π^2 , where any two "neighbouring"
vertices are connected by an edge , is closely related to the
questions discussed above. In /L.BlSa/ it is mentioned that the
discrete plane can be searched by two cooperating automata using
one pebble, or by a 3-pebble automaton , but not by a 2-pebble automaton. In /L.Sz83a,b/ and /L.And/ , one finds some further upper
and lower bounds of the machine complexity of searching the discrete plane. Note that the automata considered in /S.Ku/ violate

the inner point of view , since they know in which quadrant of the
plane they are acting , for any step.

4.4. On some other results and problems

Throughout this book we have preferably considered searching
problems on our basic types of labyrinths. These indeed represent
the core of labyrinth theory , since they are also connected with
most of the other problems in this area.

At the end of Section 1.6 , we have shown the equivalence of
certain searching and mastering problems,respectively. The reader
is surely able to find some further analogous relationships , and
straightforwardly one can show the equivalence of some searching
problems to escaping problems on corresponding types of open
labyrinths.

There is a close connection between searching problems and the
decision or recognition of properties of labyrinths. Let $\mathcal{L}$ be a
type of labyrinths and , as in Section 1.6 ,

$$\hat{\mathcal{L}} = \{ (L,h) : L \in \mathcal{L} \text{ and } h \text{ is a position in } L \}.$$

By a <u>locally determined property</u> (of depth k), we mean a mapping

$$P : \hat{\mathcal{L}} \longrightarrow \{ \text{TRUE , FALSE} \}$$

such that

$$P (L_1 , h_1) = P (L_2 , h_2)$$

if the k-neighbourhood of h_1 in L_1 (that is the corresponding
sublabyrinth) is isomorphic to the k-neighbourhood of h_2 in L_2 ,
where h_2 is the image of h_1 .

Let , for $L \in \mathcal{L}$,

$$P_{ex} (L) = \begin{cases} \text{TRUE if there is a position } h \text{ in } L \text{ with} \\ \qquad\qquad\qquad\qquad\qquad P(L,h) = \text{TRUE} , \\ \text{FALSE} \quad \text{otherwise} . \end{cases}$$

<u>Supplement 9.</u> Let $\mathcal{L}$ be a type of labyrinths , $\mathcal{A}$ a corresponding
type of automata and P a locally determined property. Under
rather weak suppositions (decidability of P by an automaton
from $\mathcal{A}$, closeness of $\mathcal{A}$ under certain compositions) , the following holds. P_{ex} can be recognized on $\mathcal{L}$ by an automaton
from $\mathcal{A}$ iff the labyrinths of $\mathcal{L}$ can be searched by an
automaton from $\mathcal{A}$. P_{ex} can be decided on $\mathcal{L}$ by an automaton
from $\mathcal{A}$ iff there is an automaton in $\mathcal{A}$ which searches all
labyrinths of $\mathcal{A}$ and always halts some time after the search.

The precise formulation for some examples one can find in
/L.Sh73,74/ and /S.HeMu/ .

For further results concerning the recognition or decision of
properties of labyrinths by several types of automata , the reader
is referred to /S.My/ , /L.Sh73,74/ and especially to the paper of
M. Ejsmont /L.Ej/.

Another kind of decision problems is treated in /L.DaKa/.
Here R. Danecki and M. Karpinski consider the problem of deciding,
for a given finite automaton , if it searches all labyrinths of a
certain type $\mathcal{L}$. They give examples of types $\mathcal{L}$ for which this
property is recursively decidable , but also examples of types on
which the property cannot be recursively decided.

In /L.PuUl/ and /L.Sz85/ the problem of rendezvous of mice
(these are finite automata) in the discrete plane is considered.
Solutions of this problem could be useful in programming cooperating systems of finite automata.

3D labyrinths are poorly considered in this book. However,one
should note their increasing importance in connection with investigations of 3-dimensional images which are produced in computed tomography. So A. Rosenfeld initiated a series of reports
on the digital geometry of 3-dimensional images , see /S.Ro81/ ,
/S.MoRo/ .

Finally , we would like to stress that the basic types of labyrinths given in Section 1.4 seem to cover the fundamental areas
of objects interesting for labyrinth theory. So the introduction
of more than four compass directions in plane C-graphoids does
not yield essentially new problems. And in /L.He87b/ we have
shown that the labyrinth problems for edge-coloured graphs (and
corresponding types of automata) are essentially equivalent to
those for R-graphs.

At this point we want to finish our travel through (the labyrinth of) labyrinth research. The author would be grateful for
any hint or comment concerning further interesting labyrinth
problems , related results or shortcomings within this book.

BIBLIOGRAPHY

L .   Labyrinth theory (in the narrower sense)

/AlKaLiLoRa/ Aleliunas , R. , R.M. Karp , R.J. Lipton , L. Lovasz ,
     C. Rackoff , Random walks ,universal traversing sequences ,
     and the complexity of maze problems. 20 th FOCS , 1979 ,
     218 - 223.

/An/   Antelmann , H. , Der Platzbedarf von Fallen für endliche
     Automaten. Dissertation A , Humboldt - Universität , Berlin
     1980.

/AnBuRo/ Antelmann , H. , L. Budach , H. - A. Rollik , On universal
     traps. EIK 15 (3) , 1979 , 123 - 131.

/And/   Анджан , А.В. , Возможности автоматов при обходе плоскости.
     Проблемы передачи информации , том XIX , 3 , 1983 ,
     78 - 89.

/As/   Asser, G. , Bemerkungen zum Labyrinth - Problem.
     EIK 13 (4/5) , 1977 , 203 - 216.

/BlKo/ Blum , M. , D. Kozen , On the power of compass. 19 th FOCS ,
     1978 , 132 - 142.

/BlSa/ Blum , M. , W.J. Sakoda , On the capability of finite au-
     tomata in 2 and 3 dimensional space. 17 th FOCS , 1977 ,
     147 - 161.

/Bu75/ Budach , L. , On the solution of the labyrinth problem for
     finite automata. EIK 11 (10 - 12) , 1975 , 661 - 672.

/Bu78a/ Будах , Л. , Автоматы в лабиринтах. Пробл. Кибернетики 34 ,
     1978 , 83 - 94.

/Bu78b/ Budach , L. , Automata and labyrinths. Math. Nachr. 86 ,
     1978 , 195 - 282.

/Bul/   Bull , M. , Fallenkonstruktion für Automaten und Automaten-
     systeme. Diplomarbeit , E.-M.-A.-Universität , Greifswald
     1986.

/BulHe/ Bull , M. , A. Hemmerling , Finite embedded trees and
     simply connected mazes cannot be searched by halting finite
     automata. Manuscript , Greifswald 1987 , to appear in EIK.

/BuMi/ Budach , L. , U. Mieth , 1-Counter automata in circles.
     Humboldt - Universität zu Berlin , Sektion Mathematik ,
     Preprint Nr. 32 , 1982.

/Co77/ Coy , W. , Automata in labyrinths. Lect. Notes in Computer
     Science 56 (Proceedings of FCT'77), 1977 , 65 - 71.

/Co78/ ---- , Of mice and maze. EIK 7 (5), 1978 , 227 - 232.

/CoRa/ Cook , S.A. , C.W. Rackoff , Space lower bounds for maze
     threadability on restricted machines. SIAM J. Comput. ,
     Vol. 9 , No. 3 , 1980 , 636 - 652.

/Da/   Danecki , R. , Finite automata and tree-like structures.
     Proc. of the Workshop on Algorithms and Computing Theory ,
     Techn. Univ. Poznan , 1981 , 20 - 23.

/DaKa/ Danecki , R. , M. Karpinski . Decidability results on plane
automata searching mazes. In: Mathematical Research v. 2 ,
Akademie – Verlag Berlin , 1979 (Proc. of FCT'79), 84 – 91.

/Do/ Döpp , K. , Automaten in Labyrinthen. EIK 7 (2 and 3),
1971 , 79 – 94 and 167 – 190.

/Ej/ Ejsmont , M. , Problems in labyrinths decidable by pebble
automata. EIK 20 (12), 1984 , 623 – 632.

/He82/ Hemmerling , A. , Zur Raumkompliziertheit von Absuchprozes-
sen auf endlichen Graphen. Rostocker Math. Kolloq. 19 ,
1982 , 77 – 90.

/He83a/ ---- , D $\neq$ ND für mehrdimensionale Turing-Automaten mit
sublogarithmischer Raumschranke. EIK 19 (1/2), 1983 ,
85 – 194.

/He83b/ ---- , Labyrinth – Probleme – eine alte Thematik mit großer
Aktualität. Math. Schülerzeitschrift "alpha" 17 , 1983 ,
73 – 75 , 100 – 102 , 129.

/He84/ ---- , A searching algorithm for finite embedded d-graphs
with no more than k regions. Bulletin of the EATCS Nr. 24 ,
October 1984 , 78 – 84.

/He85/ ---- , On labyrinth problems. In: Graphs ,Hypergraphs and
Applications. Teubner – Texte zur Mathematik Bd. 73 , 1985 ,
57 – 61.

/He86a/ ---- , 1-Pointer automata searching finite plane graphs.
Zeitschr. f. math. Logik u. Grundlagen d. Math. 32 , 1986 ,
245 – 256.

/He86b/ ---- , Remark on the power of compass. Lect. Notes in
Comp. Sc. 233 (Proc. MFCS'86), 1986 , 405 – 413.

/He87a/ ---- , Normed two-plane traps for finite systems of cooperating compass automata. EIK 23 (8/9), 1987 , 453 – 470.

/He87b/ ---- , Edge-coloured graphs in labyrinth theory (short
outline). Vortragsauszüge "Kombinatorik und Anwendungen" ,
E.-M.-A.-Universität Greifswald , Preprint-Reihe Mathematik
18 , 1987 , 24 – 27.

/He87c/ ---- , Three-dimensional traps and barrages for cooperating automata – extended abstract. Lect. Notes in Comp. Sc.
278 (Proc. FCT'87), 1987 , 197 – 203.

/He88/ ---- , Resultate und offene Probleme der Labyrinth – For-
schung. Wiss. Zeitschr. PH Güstrow , Mathem.-Naturw.
Fakultät , Nr. 1 / 1988 , 63 – 74.

/HeKr/ Hemmerling , A. , K. Kriegel , On searching of special
classes of mazes and finite embedded graphs. Lect. Notes
in Comp. Sc. 176 (Proc. MFCS'84), 1984 , 291 – 300.

/Ho81/ Hoffmann , F. , One pebble does not suffice to search planar
labyrinths. Lect. Notes in Comp. Sc. 117 (Proc. FCT'81),
1981 , 433 – 444.

/Ho82/ ---- , 1-Kiesel – Automaten in Labyrinthen. Report R-Math-
06 / 82 , AdW der DDR , Berlin 1982.

/Ho84 a,b/ ---- , Four pebbles don't suffice to search planar
infinite labyrinths.
Preprint P – MATH – 40 / 84 , AdW der DDR , Berlin 1984
and
Colloquia Mathematica societatis Janos Bolyai , 44. Theory
of Algorithms , Pecs(Hungary) , 1984 , 191 – 206.

/HoKr/ Hoffmann , F. , K. Kriegel , Quasiplane labyrinths.
Preprint P-MATH-20/83 , AdW der DDR , Berlin 1983.

/KaPaSi/ Karloff , H.J. , R. Paturi , J. Simon , Universal traversal sequences of length $n^{O(\log n)}$ for cliques.
Information Processing Letters 28 , 1988 , 241 - 243.

/Ko/ König , D. , Theorie der endlichen und unendlichen Graphen.
Akademische Verlagsgesellschaft M.B.H. , Leipzig 1936;
new edition (Mit einer Abhandlung von L. Euler) :
Teubner - Archiv zur Mathematik Bd. 6 , Leipzig 1986.

/KoTh/ Koegst , M. , K. Thalwitzer , Zum Labyrinthproblem. EIK 3
(6) , 341 - 350.

/Koz/ Kozen , D. , Automata and planar graphs. In: Mathematical
Research v. 2 , Akademie-Verlag Berlin (Proc. FCT'79),
1979 , 243 - 254.

/Kr/ Kriegel , K. , Universelle 1-Kiesel-Automaten für k-kompo-
nentige Labyrinthe. Report R-MATH - 04 / 84 , AdW der DDR ,
Berlin 1984.

/Mu71/ Müller , H. , Endliche Automaten und Labyrinthe. EIK 7 (4),
1971 , 261 - 264.

/Mu77 a/ ---- , A one-symbol printing automaton escaping from
every labyrinth. Computing 19 , 1977 , 95 - 110.

/Mu77 b/ ---- , Automaten in Labyrinthen. In: Beiträge zur Theorie
der Polyautomaten. TU Braunschweig , Informatik - Berichte
(Preprint) , 1977 , 68 - 76.

/Mu79/ ---- , Automata catching labyrinths with at most three
components. EIK 15 (1/2), 1979 , 3 - 9.

/PuUl/ Pultr , A. , J. Ulehla , Rendezvous of mice. Proc. of the
Workshop on Algorithms and Computing Theory , Techn. Univ.
Poznan , 1981 , 43 - 48.

/Ro/ Rollik , H. - A. , Automaten in planaren Graphen.
Lect. Notes in Comp. Sc. 67 (Proc. of the 4 th GI Conf. on
TCS) , 1979 , 266 - 275
and
Acta Informatica 13 , 1980 , 287 - 298.

/Ros/ Rosenstiehl , P. , Labyrinthologie mathématique I. Math.
Sci. hum. 9 (33), 1971 , 5 - 32.

/Sc/ Schreiber , P. , Theseus im Labyrinth als Turingmaschine.
Zeitschr. f. Math. Logik u. Grundlagen d. Math. 17 , 1971 ,
57 - 60.

/Sh73/ Shah , A.N. , On traversing properties of array automata.
Univ. of Maryland , Computer Science Center , TR - 274 , 1973.

/Sh74/ ---- , Pebble automata on arrays. Comp. Graph. Image Proc.
3 , 1974 , 236 - 246.

/Shan/ Shannon , C.E. , Presentation of a maze-solving machine.
In: Cybernetics , Trans. of the 8 th Conf. of the Josiah
Macy Jr. Found. , 1951 , 173 - 180.

/Sz82/ Szepietowski , A. , A finite 5-pebble automaton can search
every maze. Information Processing Letters 15 (5), 1982 ,
199 - 204.

/Sz83 a/ ---- , Remarks on searching labyrinths by automata.
Lect. Notes in Comp. Sc. 158 (Proc. FCT'83), 1983 ,
457 - 464.

/Sz83 b/ ---- , On searching plane labyrinths by 1-pebble automa-
         ta. EIK 19 (1/2), 1983 , 79 - 84.

/Sz85/   ---- , On Paterson's problem. EIK 21 (6), 1985 , 313 - 314.

/Ul/   Ulehla , J. , On racial insufficiency: white messengers
         cannot simulate coloured ones. Commentationes mathematicae
         universitatis Carolinae 26 (2), 1985 , 323 - 325.

/Wi/   Wiener , C. , Ueber eine Aufgabe aus der Geometria situs.
         Mathematische Annalen 6 , 1873 , 29 - 30.

S. Surroundings (applications , motivations , graphs , complexity,
               and other more or less related topics)

/AhHoUl/ Aho , A.V. , J.E. Hopcroft , J.T. Ullman , The design and
         analysis of computer algorithms. Addison - Wesley P.C. ,
         Reading (Mass.) 1974.

/Be/   Berge , C. , Graphs and hypergraphs. North-Holland P.C. ,
         Amsterdam and London 1973.

/BlHe/   Blum , M. , C. Hewitt , Automata on a 2-dimensional tape.
         8 th IEEE Conf. on SWAT , 1967 , 155 - 160.

/Bu77/   Budach , L. , Environments , labyrinths and automata. Lect.
         Notes in Comp. Sc. 56 (Proc. FCT'77), 1977 , 54 - 64.

/Bu81/   ---- , Two pebbles don't suffice. Lect. Notes in Comp. Sc.
         118 (Proc. MFCS'81), 1981 , 578 - 589.

/BuMe/   Budach , L. , C. Meinel , Environments and automata. EIK 18
         (1 / 2 and 3), 1982 , 3 - 40 and 115 - 139.

/Fa/   Fary , I. , On straight line representation of planar
         graphs. Acta Scientiarum Mathematicarum Szeged Tomus XI ,
         4 , 1948 , 229 - 233.

/Fi/   Fischer , P.C. , Turing machines with restricted memory
         access. Information and Control 9 , 1966 , 364 - 379.

/GaJo/   Garey , M.R. , D.S. Johnson , Computers and intractability :
         a guide to the theory of NP-completeness. Freeman , San
         Francisco 1979 .

/Ha/   Harary , F. , Graph theory. Addison - Wesley P.C. , Reading
         1969.

/He79 a/ Hemmerling , A. , Concentration of multidimensional tape-
         bounded systems of Turing automata and cellular spaces.
         In: Mathematical Research v. 2 , Akademie-Verlag Berlin ,
         1979 (Proc. FCT'79), 167 - 174.

/He79 b/ ---- , Zur Raumkompliziertheit mehrdimensionaler Zellu-
         larräume und Turing-Automaten. EIK 15 (3), 1979 , 143 - 158

/He86/   ---- , On the power of cellular parallelism (ext. abstr.).
         In: Mathematical Research v. 29 , Akademie-Verlag Berlin ,
         1986 (Proc. of PARCELLA'86), 210 - 217.

/HeMu/   Hemmerling , A. , G. Murawski , Zur Raumkompliziertheit mehr-
         dimensionaler Turing-Automaten. Zeitschr. f. math. Logik
         u. Grundlagen d. Math. 30 , 1984 , 233 - 258.

/Ho/   Horowitz , E. , Fundamentals of programming languages (2nd
         Edition), Springer-Verlag , Berlin et al., 1984 .

/HoKr/ Hoffmann , F. , K. Kriegel , Embedding rectilinear graphs in
linear time. Information Processing Letters 29 , 1988 ,
75 - 79.

/HoSa/ Horowitz , E. , S. Sahni , Algorithmen . Entwurf und Analyse.
Springer-Verlag , Berlin et al., 1981.

/KoTr/ Kobrinski , N.E. , B.A. Trachtenbrot , Einführung in die
Theorie endlicher Automaten. Akademie-Verlag , Berlin , 1967.

/Ku/ Курдюмов, Г.Л., Коллектив автоматов с универсальной про-
ходимостью. Проблемы передачи информации, том XVII , 4 ,
I98I , 98 - II2.

/KuAlPo/ Кудрявцев, В.Б. , С.В. Алёшин , А.С. Подколзин , Элементы
теории автоматов. изд. МГУ, Москва I978.

/La/ Landau , E. , Handbuch der Lehre von der Verteilung der
Primzahlen,1. Bd. . B.G. Teubner , Leipzig u. Berlin 1909.

/LeSe/ Leong , B.L. , J.I. Seiferas , New real-time simulations of
multihead tape units. Journal of the ACM , v. 28 , 1981 ,
166 - 180.

/Ma/ Matthews , W.H. , Mazes and labyrinths : their history and
development. New York 1970.

/Me/ Meinel , C. , The importance of plane labyrinths. EIK 18
(7/8) , 1982 , 419 - 422.

/Mi/ Minsky , M.L. , Computation : finite and infinite machines.
Prentice - Hall , Inc. Englewood Cliffs , N.J. , 1967.

/MiRo/ Milgram , D.L. , A. Rosenfeld , Array automata and array
grammars. Information Processing 71 (IFIP'71) Conf. Proc.,
North-Holland , Amsterdam , 1972 , 69 - 74.

/MoRo/ Morgenthaler , D.G. , A. Rosenfeld , Surfaces in three-dimensional digital images. Information and Control 51 , 1981 ,
227 - 247.

/Mu/ Müller , H. , Stackautomaten in Labyrinthen. Archiv f. math.
Logik u. Grundlagenforschung 14 , 1971 , 127 - 134.

/My/ Mylopoulos , J. , On the recognition of topological in-
variants by 4-way finite automata. Computergraphics and
Image Processing 1 , 1972 , 308 - 316 .

/Or/ Ore , O. , Theory of graphs. Am. Math. Soc. , Providence
1962.

/PrSh/ Preparata , F.P. , M.I. Shamos , Computational geometry.
Springer-Verlag , Heidelberg,New York 1984.

/Ro70/ Rosenfeld , A. , Connectivity in digital pictures. J.ACM
v. 17 (1), 1970 , 146 - 160.

/Ro76/ ---- , Array and web grammars : an overview. In: Automata,
Languages , Development , North-Holland 1976 , 517 - 529.

/Ro81/ ---- , Three-dimensional digital topology. Information and
Control 50 , 1981 , 119 - 127.

/RoFiHo/ Rosenstiehl , P. , J.R. Fiksel , A. Holliger , Intelligent
graphs : networks of finite automata capable of solving
graph problems. In: Graph theory and computing , Academic
Press , New York and London 1972 , 219 - 265.

/Sa/ Sachs , H. , Einführung in die Theorie der endlichen Graphen,
I and II . B.G. Teubner , Leipzig 1970 and 1972 , resp.

/Sal/ Salomaa , A. , Formal languages. Academic Press , New York
and London 1973.

/SaSt/ Sack , J.-R. , T. Strothotte , Capturing winding information
        of polygons and its implications for the design of effi-
        cient algorithms. Manuscript.

/Sav70/ Savitch , W. , Relationships between nondeterministic and
        deterministic tape complexities. JCSS 4 (2), 1970 , 177-192.

/Sav73/ ---- , Maze recognizing automata and nondeterministic
        tape complexity. JCSS 7 (4), 1973 , 389 - 403 .

/St/   Stahl , S. , The embeddings of a graph - a survey. J. of
        graph theory 2 , 1978 , 275 - 298 .

/Ste/   Stein , S.K. , Convex maps. Proc. Amer. Math. Soc. 2 , 1951 ,
        464 - 466 .

/Th/   Thomassen , C. , Straight line representations of infinite
        planar graphs. J. London Math. Soc. 16 , 1977 , 411 - 423 .

/ViWi/ Vijayan , G. , A. Wigderson , Rectilinear graphs and their
        embeddings. SIAM J. Comput. v. 14 (2), 1985 , 355 - 372 .

/Wa/   Wagner , K. , Bemerkungen zum Vierfarbenproblem. Jahresbe-
        richt der Deutschen Mathematikervereinigung Bd. 46 , 1936 ,
        26 - 32 .

/WaWe/ Wagner , K. , G. Wechsung , Computational complexity. VEB
        Deutscher Verlag der Wissenschaften , Berlin 1986 .

/Wi/   Wiedermann , J. , Searching algorithms. Teubner-Texte zur
        Mathematik Bd. 99 , Leipzig 1987 .

/Yo/   Youngs , J.W.T. , Minimal imbeddings and the genus of a
        graph. J. Math. Mech. 12 , 1963 , 303 - 316 .

Q. Quotations (at the heads of the introduction and chapters)

O .   Hermann Hesse , Das Glasperlenspiel.
      Aufbau - Verlag , Berlin und Weimar 1985 , p. 16 .

I .   Wilhelm Busch , Zwiefach sind die Phantasien.
      Verlag Philipp Reclam jun. , Leipzig 1972 , p. 337.

II . Umberto Eco , Der Name der Rose.
      Verlag Volk und Welt , Berlin 1985 , p. 216 .

III . Ovid , Werke in zwei Bänden. Erster Band: Verwandlungen.
      Aufbau - Verlag , Berlin und Weimar 1973 , p. 187.

IV . Jorge Luis Borges , Labyrinthe.
      Deutscher Taschenbuch - Verlag , München 1962 , pp. 35 , 36.

<u>**Some predefined symbols**</u>

$\in , \subseteq , \cup , \bigcup , \cap , \bigcap , \smallsetminus , \times , X$	set theoretic relations , operations
$\emptyset$	empty set
card(S)	cardinality of the set S
max(S) , min(S)	maximum , minimum of the set S
$\mathbb{N} , \mathbb{N}^+$	set of all (positive) natural numbers
$\mathbb{Z}$	set of all integers
$\mathbb{R} , \mathbb{R}^+ , \mathbb{R}^-$	set of all (positive,negative) real numbers
$\|z\|$	absolute value , euclidean norm of z
$[x , y]$	closed interval in $\mathbb{R}$
X^*	set of all words over the alphabet X
Λ	empty word
X^+	$= X^* \smallsetminus \{ \Lambda \}$
len(w)	length of the word w
$w_1 \cdot w_2 , w_1 w_2$	concatenation of words
w^k	k th power of the word w
w^∞	infinite power (yields a sequence)
$X^{\mathbb{N}}$	set of all (infinite) sequences over X
X^k	k th cartesian power of X , or $\{ w \in X^* : \text{len}(w) = k \}$ (context-dependent)
$f : S_1 \longrightarrow S_2$	f is a mapping of S_1 into S_2
$f(s) , f(S)$	image of $s \in S_1$, complete image of $S \subseteq S_1$
$f_{/S}$	restriction of f to $S \subseteq S_1$
$f \circ g$	product of mappings (f after g)
f^k	k th power of the mapping f

L. A. Kalužnin / P. M. Beleckij / V. Z. Fejnberg

<u>Kranzprodukte</u>

Das vorliegende Buch behandelt Kranzprodukte von Permutationsgruppen
(und Transformationsgruppen) und unterscheidet sich damit von vielen
anderen Publikationen, in denen Kranzprodukte anderer algebraischer
Strukturen (z.B. abstrakter Gruppen oder Halbgruppen) betrachtet werden. Kranzprodukte von Permutationsgruppen wurden unter dem Namen
"produit complêt" von L. A. Kaloujnine (= Kalužnin) und M. I. Krasner
in den 40er Jahren eingeführt. Historisch kann man den Begriff des
Kranzproduktes bis in das 19. Jahrhundert zurückverfolgen. Anwendungen
des Kranzproduktes für Permutationsgruppen gibt es in der mathematischen Chemie und der Informatik. Innerhalb der Mathematik finden Kranzprodukte Anwendung besonders in der abstrakten Gruppentheorie wie auch
in der Theorie der Permutationsgruppen und führten zu wichtigen Ergeb-
nissen in beiden Theorien (Schreiersches Gruppenerweiterungsproblem,
Geometrie ultrametrischer Räume). Das Buch wendet sich vor allem an
Studenten und Hochschullehrer auf dem Gebiet der reinen und angewand-
ten Mathematik, besonders der Informatik, aber auch an Mathematiker
und Informatiker mit algebraischen Interessen.
Bd. 101, 167 S., DDR 17,50 M, Ausland 17,50 DM,
ISBN 3-322-00425-2

G. Schaar / M. Sonntag / H.-M. Teichert

<u>Hamiltonian Properties of Products of Graphs and Digraphs</u>

This book gives a survey on the main results concerning the subject described by the title, also considering the contributions made by the
authors in this field. The central object is to study the dependence
of the Hamiltonian behaviour of given products of graphs on properties
of the factors. Moreover, the classical products (Cartesian sum, lexicographic product, disjunction, Cartesian product, normal product) are
particularly investigated in connection with such Hamiltonian properties as traceability, Hamiltonicity, higher Hamiltonicity, Hamiltonian
connectedness, strong path-connectedness, pancyclicity, decomposability
into Hamiltonian cycles. The parallel treatment of this set of problems
for undirected and directed graphs provides the possibility of a com-
parative consideration with regard to similarities and differences.
Bd. 108, 148 S., 1988, DDR 15,50 M, Ausland 15,50 DM,
ISBN 3-322-00501-1

<u>Seminar Analysis of the Karl-Weierstraß-Institute 1986/87</u>

Ed. by B.-W. Schulze and H. Triebel

The Teubner-Text 'Seminar Analysis' is the continuation of a corresponding series published by the Karl-Weierstraß-Institute of Mathematics of the Academy of Sciences of the GDR 1981 - 1985. The volume
1985/86 appeared as the Teubner-Text 96. The main aim of this series
is the publication of survey papers on modern analysis, in particular
functional analytic and structure methods in partial differential equations, complex function theory, mathematical physics, global analysis
and differential geometry. Another part contains short announcements of
outstanding results on these subjects. The present volume contains ar-
ticles on elliptic operators on non-compact manifolds, in particular
with conical singularities, global analysis, infinite-dimensional super-
manifolds, functional analysis, propagation of singularities.
Bd. 106, 332 S., 1988, DDR 34,50 M, Ausland 34,50 DM,
ISBN 3-322-00503-8

Diese Reihe wurde geschaffen, um eine schnellere Veröffentlichung
mathematischer Forschungsergebnisse und eine weitere Verbreitung von
mathematischen Spezialvorlesungen zu erreichen. TEUBNER-TEXTE werden
in deutsch, englisch, russisch oder französisch erscheinen. Um Aktualität der Reihe zu erhalten, werden die TEUBNER-TEXTE im Manuskript-
druck hergestellt, da so die geringeren drucktechnischen Ansprüche
eine raschere Herstellung ermöglichen. Autoren von TEUBNER-TEXTEN
liefern an den Verlag ein reproduktionsfähiges Manuskript. Nähere
Auskünfte darüber erhalten die Autoren vom Verlag.

This series has been initiated with a view to quicker publication of
the results of mathematical research-work and a widespread circula-
tion of special lectures on mathematics. TEUBNER-TEXTE will be published in German, English, Russian or French. In order to keep this
series constantly up to date and to assure a quick distribution, the
copies of these texts are produced by a photographic process (small-
offset printing) because its technical simplicity is ideally suited
to this type of publication. Authors supply the publishers with a
manuscript ready for reproduction in accordance with the latter's
instructions.

Cette série des textes a été créée pour obtenir une publication plus
rapide de résultats de recherches mathématiques et de conférences
sur des problèmes mathématiques spéciaux. Les TEUBNER-TEXTE seront
publiés en langues allemande, anglaise, russe ou française. L'actua-
lité des TEUBNER-TEXTE est assurée par un procédé photographique
(impression offset). Les auteurs des TEUBNER-TEXTE sont priés de
fournir à notre maison d'édition un manuscrit prêt à être reproduit.
Des renseignements plus précis sur la form du manuscrit leur sont
donnés par notre maison.

Эта серия была создана для обеспечения более быстрого публикования
результатов математических исследований и более широкого распростра
нения математических лекций на специальные темы. Издания серии
ТОЙБНЕР-ТЕКСТЕ будут публиковаться на немецком, английском, русском
или француском языках. Для обеспечения актуальности серии, ее издания будут изготовляться фотомеханическом способом. Таким образом,
более скромные требования к полиграфическому оформлению обеспечат
более быстрое появление в свет. Авторы изданий серии ТОЙБНЕР-ТЕКСТЕ
будут предоставлять издательству рукописи, удовлетворящие требова-
ниям фотомеханического печатания. Более подробные сведения авторы
получат от издательства.

BSB B. G. Teubner Verlagsgesellschaft, Leipzig
DDR - 7010 Leipzig, Postfach 930